Performance Analysis and Optimization of Multi-Traffic on Communication Networks

Leonid Ponomarenko · Che Soong Kim ·
Agassi Melikov

Performance Analysis and Optimization of Multi-Traffic on Communication Networks

Leonid Ponomarenko
National Academy of Science of Ukraine
International Scientific-Education Center
Melnikova str 49, ap. 32
04050 Kiev
Ukraine
laponomarenko@ukr.net

Che Soong Kim
Sangji University
Department of Industrial Engineering
220-702 Wonju, Kangwon
Korea, Republic of Korea
dowoo@sangji.ac.kr

Agassi Melikov
National Academy of Sciences
Department of Teletraffic
Institute of Cybernetics
F.Agayev str. 9
AZ1141 Baku
Azerbaijan
agassi.melikov@rambler.ru

ISBN 978-3-642-42340-6 ISBN 978-3-642-15458-4 (eBook)
DOI 10.1007/978-3-642-15458-4
Springer Heidelberg Dordrecht London New York

Softcover re-print of the Hardcover 1st edition 2010

Cover design: WMXDesign GmbH, Heidelberg

Printed on acid-free paper

Springer is part of Springer Science+Business Media (www.springer.com)

Preface

Rapidly growing demand for telecommunication services and information interchange has led to communication becoming one of the most dynamic branches of the infrastructure of modern society. On the boundary between the twentieth and twenty-first century the convergence of telecommunication, computer and information technologies and also the merging of corresponding industries begun and is proceeding.

The teletraffic theory has become in the last few years an independent scientific discipline representing a set of probabilistic methods to solve problems of planning and optimization of telecommunication systems. For the solution of problems that arise in the specified branch of engineering practice both analytical and numerical methods of modeling are widely used. However, the application areas of analytical methods are essentially limited due to the complexity of models of modern teletraffic systems. Therefore, in modern teletraffic theory there is a special urgency for the development of effective numerical methods of research. Specific features of various teletraffic systems aggravate problems of development of universal numerical methods.

The functioning of complex probabilistic systems is described by mathematical models considering the basic specific features of the work of these systems. In such mathematical models various classes of stochastic processes are used.

The most effective mathematical tool for analysis has been developed for systems in which functioning is described by homogeneous Markov processes (or chains). The basic characteristics of Markov models are determined by solution of systems of linear equations (algebraic, differential, or integral). However, the assumption concerning the Markov property of investigated systems is rather limiting, therefore as mathematical models of real systems more general classes of stochastic processes are used. At the same time, sometimes it is possible to achieve Markov property for a model by complicating state space of the model process.

The tendency for maximal accuracy in the description of functioning of real systems leads to the fact that corresponding mathematical models have become more and more complex. Thus, their mathematical analysis becomes complicated and the tool for the analysis becomes cumbersome and often inaccessible in an engineering practice.

The basic difficulty in modeling and analysis of complex telecommunication systems is shown by the unreasonable increase in the number of the various states which leads to the practical immensity of a model. An actual problem in the mathematical theory of teletraffic is overcoming a basic difficulty – the large dimension of state space of the model ("curse of dimensionality").

Here the problem consists of using the large number of possible states of a system as a positive factor essentially allowing simplification of its analysis (similarly to statistical physics where the basic laws of large-scale physical systems represent the interaction of the ultimately huge number of particles forming them).

The most radical approach to overcoming complexity in the analysis of a real system consists of construction of a simpler merged system whose analysis is essentially easier than the real analysis, and the basic characteristics can be accepted as characteristics of the latter.

The idea of researching a complex system via its parts at the level of subsystems with a subsequent transition to the system underlies methods of solution of various problems. Corresponding methods are developed in essence to find solutions for large systems of algebraic equations. Thus, the main point here is the fact that the final solution of the system in its parts appears to be the exact solution of the initial system of equations. However, the mentioned methods which have undoubtedly proved their importance, appear in many cases to be insufficient for the solution of problems arising in an engineering practice.

In the present book methods for the merging of state space of complex teletraffic systems are uniformly developed. The idea of these methods is that the state space of a real system is split into a finite or infinite number of disjoint classes. The states of each of these classes are merged into one state. In the new merged state space a merged system whose functioning in a certain sense describes the functioning of the initial system is under construction.

The developed numerical methods can be applied to research of both classical models of integrated voice/data cellular wireless networks (CWN) and multi-rate models of multimedia communication networks with an arbitrary number of heterogeneous calls. Thus, the efficiency of the offered methods is shown on concrete classes of teletraffic systems. It is important to note, that they can be successfully applied, also in other branches of engineering practice.

The book consists of two parts and an appendix. In Part I which consists of three chapters, the models of a CWN integrating voice and data calls are investigated. Unlike classical models here it is supposed that original calls and handover (handoff) calls can differ from each other in terms of time of occupation of radio channels of a cell. In Chap. 1 two-dimensional unbuffered models of the investigated CWN are examined using various multi-parametric call admission control (CAC) strategies. In Chap. 2 similar models are investigated in the presence of finite and infinite queues of heterogeneous calls, thus handover calls can have limited sojourn distribution in a cell. In the last chapter of this part, problems of improvement of Quality of Service (QoS) metrics are considered.

In Part II which consist of four chapters, the multi-rate queuing models of multimedia communication networks intended for transfer of diverse messages are investigated. In Chap. 4 effective algorithms for calculation of QoS metrics in such networks are developed for various CAC strategies. Thus it is supposed, that all calls are inelastic. In Chap. 5 mixed models in which joint service of inelastic and elastic calls is carried out are investigated. Thus, models of two types are examined: with discrete and continuous schemes of distribution of a bandwidth between elastic calls. In Chap. 6 the problems of parametric optimization of QoS metrics of heterogeneous calls in multimedia communication networks are investigated. In Chap. 7 problems of the application of methods of the Markov Decision Processes (MDP) in teletraffic systems are considered. First a general scheme of the application of MDP in investigated systems is shown and the exact and approximate methods of solution of corresponding optimization problems are described. Then problems of finding an optimal and sub-optimal CAC strategy in models of multimedia communication networks are described, thus here a criterion of efficiency is maximization of channel utilization.

In the appendix a new approximate method for calculation of models of teletraffic systems which are described by a two-dimensional Markov chain (2-D MC) is developed.

The reader should know the basics of the theory of telecommunications, queuing theory, numerical methods, and methods of mathematical programming at the graduate or advanced undergraduate level.

The book contains the original results of authors published over the last few years in known scientific journals and reported at representative international scientific conferences. Each chapter is accompanied by a comment and a reference list which when combined gives an almost complete representation of the modern state of research of the mathematical theory of teletraffic. The book will be useful to specialists in the field of telecommunication systems and also to students and post-graduate students of corresponding specialties.

Kiev, Ukraine — Leonid Ponomarenko
Gangwon-do, Korea — Che Soong Kim
Baku, Azerbaijan — Agassi Melikov

[illegible] general [illegible]

[illegible] [illegible]
[illegible] [illegible]
[illegible]

Contents

List of Acronyms and Abbreviations

BDP	Birth-and-Death Process;
BS	Base Station;
CAC	Call Admission Control;
CS	Complete Sharing;
CSE	Complete Sharing with Equalization;
CWN	Cellular Wireless Network;
FIFO	First-In-First-Out;
GC	Guard Channels;
GM	Generating Matrix;
MC	Markov Chain;
MDP	Markov Decision Process;
MRQ	Multi-Rate Queue;
MS	Mobile Subscriber;
MSC	Mobile Switching Center;
PB	Probability of Blocking;
PMA	Phase-Merging Algorithm;
QoS	Quality of Service;
SGBE	System of Global Balance Equations;
SGC	Special Group of Channels;
SLBE	System of Local Balance Equations;
TR	Trunk Reservation;
$I(A)$	Indicator function of the event A;
$P(A)$	Probability of the event A;
$\varnothing$	Empty set;
$\approx$	Approximate Equality;
$a := b$	a is defined by expression b;
$i = \overline{1,N}$	$1 \leq i \leq N$ or $i = 1, \ldots, N$;
$Int(x)$	Integer (or whole) part of x; $x^+ := max(0,x)$;
h-call	handover call;
hd-call	handover data call;
hv-call	handover voice call;
i-call	call of type i or call from the traffic of type i;

n-call	narrow-band call;
o-call	original call;
od-call	original data call;
ov-call	original voice call;
w-call	wide-band call;
$\lambda_h(\mu_h)$	arrival (service) rate of handover calls;
$\lambda_i(\mu_i)$	arrival (service) rate of type *i* calls;
$\lambda_n(\mu_n)$	arrival (service) rate of narrow-band calls;
$\lambda_o(\mu_o)$	arrival (service) rate of new calls;
$\lambda_w(\mu_w)$	arrival (service) rate of wide-band calls;
$v_h = \lambda_h/\mu_h$	load offered by h-calls;
$v_i = \lambda_i/\mu_i$	load offered by calls of type *i*;
$v_n = \lambda_n/\mu_n$	load offered by n-calls;
$v_o = \lambda_o/\mu_o$	load offered by o-calls;
$v_w = \lambda_w/\mu_w$	load offered by w-calls;
P_o	Probability of blocking of o-calls;
P_h	Probability of blocking (or dropping) of h-calls;
PB_i	Probability of blocking of calls of type *i*;
PB_n	Probability of blocking of n-calls;
PB_w	Probability of blocking of w-calls;
$p(x), \pi(x), \rho(x)$	Stationary probability of state *x*;
$x \rightarrow y$	Transition from state *x* to state *y*;
$q(x, y)$	Transition intensity from state *x* to state *y*;
$\delta_{i,j} = \begin{cases} 1, & \text{if } x > 0, \\ 0, & \text{if } x \leq 0; \end{cases}$	Kronecker's symbols;
$\delta_{i,j} = \begin{cases} 1, & \text{if } i = j, \\ 0, & \text{if } i \neq j \end{cases}$	
$\theta_j(v,m)$	Stationary distribution of Erlang's model *M/M/m/0* with load of *v* Erl, $j = 0,1,\ldots,m$;
$E_B(v,m)$	Erlang's B-formula for the model *M/M/m/0* with load of *v* Erl;
$\boldsymbol{e}_i$	unit vector in direction *i* in a Euclidean space whose dimension is specified in each particular case.

Part I
Two-Dimensional Traffic Models of Integrated Voice/Data Cellular Wireless Networks

In this part of the book models of integrated voice/data cellular wireless networks (CWN) are researched. Nowadays such networks serve more than a billion users all over the world. Researching them from a methodological point of view is also important, since researches on multimedia networks with an arbitrary number of heterogeneous calls are based on research results of such networks.

The technology of the given networks is explored in detail in special literature [2, 18, 20, 21, 28–32, 35]. Mathematical theories of the networks noted above are also given in monographic materials [1, 6, 13, 38]. In these works as well as in reviews [7, 14, 33] detailed information about known queuing models of the given networks and methods of calculation of their quality of service (QoS) metrics are available.

A wireless cellular network consists of radio access points, called base stations (BS), each covering a certain geographic area. With distance the power of radio signals fade away (fading or attenuation of signal occurs) which makes it possible to use the same frequencies over several cells, but in order to avoid interference, this process must be carefully planned. For better use of frequency recourse, existing carrier frequencies are grouped, and the number of cells, in which this group of frequencies is used, defines the so-called frequency reuse factor. Therefore, in densely populated areas with a large number of mobile subscribers (MS) small-dimensioned cells (microcells and picocells) are to be used, because of limitations of volumes and frequency reuse factor.

In connection with the limitation of transmission spectrum in wireless networks, problems of allocation of a common spectrum among cells are very important. The unit of the wireless spectrum, necessary for serving a single user is called a channel (for instance, time slots in TDMA are considered to be channels).

There are three solutions for the channel allocation problem: Fixed Channel Allocation (FCA), Dynamic Channel Allocation (DCA), and Hybrid Channel Allocation (HCA). The advantages and disadvantages of each of these are well known. At the same time, owing to simplicity of realization, the FCA scheme is widely used in existing cellular networks. In this part of the book wireless network models with FCA schemes are examined.

Quality of service in a certain cell with the FCA scheme could be improved if rational call admission control (CAC) strategies for heterogeneous traffic are provided. The use of such an access strategy doesn't require much in the way of resources, therefore this method could be considered an operative and more defensible solution of the problem of resource shortages.

Apart from original (or new) call (o-call) flows an additional class of calls that requires a special approach also exists in wireless cellular networks. These are so-called handover calls (h-calls). This is specific only for wireless cellular networks.

The essence of this phenomenon is that moving MS, that have already established a connection with the network, pass boundaries between cells and are then served by a new cell. From a new cell's point of view this is an h-call, and since the connection with the MS has already been established, MS handling transfer to a new cell must be transparent for the user. In other words, in wireless networks the call may occupy channels from different cells several times during a call's duration, which means that channel occupation period is not the same as call duration.

Mathematical models of call-handling processes in integrated voice/data cellular wireless networks can be developed adequately enough based on the theory of networks of queues with different types of calls and random topology. Such models are researched poorly in the literature, for example see [3, 4, 19]. This is explained by the fact that despite the elegance of those models, in practice they are useful only for small-dimensional networks and with some limiting simplifying assumptions that are contrary to the fact in real functioning wireless networks. Because of this, in the majority of research works models of an isolated cell are analyzed.

In almost all available works one-dimensional queuing models of call-handling processes in an isolated cell of a mono-service CWN are proposed. However, these models can not describe the studied processes since in integrated voice/data CWN heterogeneous traffic calls are quite different with respect to their bandwidth requirements, arrival rates, and channel occupancy time. Related to this in this part of the book new two-dimensional queuing models of investigated networks are developed.

Chapter 1
Performance Analysis of Multi-Parametric Call Admission Control (CAC) Strategies in Unbuffered Cellular Wireless Networks

In integrated voice/data Cellular Wireless Networks (CWN) voice calls (v-calls) are more susceptible to possible losses and delays than data (original or handover) calls (d-calls). That is why a number of different call admission control (CAC) strategies for prioritization of v-calls are suggested in various works, mostly implying the use of guard channels for high-priority calls and/or cutoff strategies which restrict the number of low-priority calls in channels.

In this chapter we introduce a unified approach to approximate performance analysis of some multi-parametric CAC strategies in a single cell of an integrated voice/data CWN which differs from known works in this area. Our approach is based on the principles of the theory of phase merging of stochastic systems [17].

With the proposed approach one can overcome an assumption made in almost all known papers about equality of handling intensities of both voice and data calls. Because of this assumption the functioning of the CWN was described with a one-dimensional Markov chain (1-D MC) and the authors presented simple formulae for calculating the QoS (Quality of Service) metrics of the system. However, as was mentioned in [38] (pages 267–268) the assumption of the same mean channel occupancy time for both original and handover calls of the same class is unrealistic. The models presented herein are more general in terms of handling intensities and the equality is no longer required.

1.1 CAC Based on the Guard Channels Strategy

It is well known that in an integrated voice/data CWN voice calls of any type (original or handover) have a higher priority than data calls and within each flow handover calls have a higher priority than original calls.

To assign priorities to handover v-calls (hv-call) in such networks a back-up scheme that involves reserving a particular number of guard channels of a cell expressly for calls of this type is a method often utilized. According to this scheme any hv-call is accepted if there exists at least one free channel, while calls of remaining kinds are accepted only when the number of busy channels does not exceed some class-dependent threshold value.

L. Ponomarenko et al., *Performance Analysis and Optimization of Multi-Traffic on Communication Networks*, DOI 10.1007/978-3-642-15458-4_1,

We consider a model of an isolated cell in an integrated voice/data CWN without queues. This cell contains N channels, $1<N<\infty$. These channels are used by Poisson flows of hv-calls, original v-calls (ov-calls), handover d-calls (hd-calls), and original d-calls (od-calls). Intensity of x-calls is λ_x, $x\in\{\text{ov, hv, od, hd}\}$. As in almost all cited works the values of handover intensities are considered known hereinafter. It is apparent that definition of their values depending on the intensity of original calls, shape of a cell, mobility of an MS (Mobile Subscriber) etc., is rather challenging and complex. However, if we consider the case of a uniform traffic distribution and at most one handover per call, the average handover intensity can be given by the ratio of the average call holding time to the average cell sojourn time [26].

For handling of any narrow-band v-call (either original or handover) only one free channel is required, while one wide-band d-call (either original or handover) requires simultaneously $b\geq 1$ channels. Here it is assumed that wide-band d-calls are inelastic, i.e. all b channels are occupied and released simultaneously (though this can be investigated for models with elastic d-calls, see Chap. 5).

Note that the channel occupancy time considers both components of occupancy time: the time of call duration and mobility. Distribution functions of channel occupancy time of heterogeneous calls are assumed to be independent and exponential, but their parameters are different, namely intensity of handling of voice (data) calls equals μ_v (μ_d), and generally speaking $\mu_\text{v}\neq\mu_\text{d}$. If during call handling a handover procedure is initiated, the remaining handling time of this call in a new cell (as an h-call) is also exponentially distributed with the same mean due to the memory-less property of exponential distribution.

In a given CAC the procedure by which the channels are engaged by calls of different types is realized in the following way. As was mentioned before, if upon arrival of an hv-call, there is at least one free channel, this call seizes one of the free channels; otherwise this call is blocked. With the purpose of definition of the proposed CAC for calls of other types three parameters N_1, N_2, and N_3 where $1\leq N_1\leq N_2\leq N_3\leq N$ are introduced. Here N_1 and N_2 are multiples of b.

A newly arrived ov-call is accepted if the number of busy channels is less than N_3, otherwise it is blocked. A newly arrived od-call (respectively, hd-call) is accepted only in the case of at most N_1-b (respectively, N_2-b) busy channels, otherwise it is blocked.

Consider the problem of finding the major QoS metrics of the given multi-parametric CAC strategy – blocking (loss) probabilities of calls of each type and overall channel utilization. For simplicity of intermediate mathematical transformations first we shall assume that $b=1$. The case $b>1$ is straightforward (see below).

By adopting an assumption for the type of distribution laws governing the incoming traffic and their holding times it becomes possible to describe the operation of an isolated cell by means of a two-dimensional Markov chain (2-D MC), i.e. in a stationary regime the state of the cell at an arbitrary moment of time is described by a 2-D vector $\boldsymbol{n}=(n_\text{d}, n_\text{v})$, where n_d (respectively, n_v) is the number of data (respectively, voice) calls in the channels. Then the state space of the corresponding Markov chain describing this call-handling scheme is defined thus:

$$S := \left\{ \boldsymbol{n} : n_{\text{d}} = \overline{0, N_2},\ n_{\text{v}} = \overline{0, N},\ n_{\text{d}} + n_{\text{v}} \le N \right\}. \tag{1.1}$$

Elements of the generating matrix of this MC $q(\boldsymbol{n}, \boldsymbol{n}'), \boldsymbol{n}, \boldsymbol{n}' \in S$ are determined from the following relations:

$$q\left(n, n'\right) = \begin{cases} \lambda_{\text{d}} & \text{if } n_{\text{d}} + n_{\text{v}} \le N_1 - 1,\ \boldsymbol{n}' = \boldsymbol{n} + \boldsymbol{e}_1, \\ \lambda_{\text{hd}} & \text{if } N_1 \le n_{\text{d}} + n_{\text{v}} \le N_2 - 1,\ \boldsymbol{n}' = \boldsymbol{n} + \boldsymbol{e}_1, \\ \lambda_{\text{v}} & \text{if } n_{\text{d}} + n_{\text{v}} \le N_3 - 1,\ \boldsymbol{n}' = \boldsymbol{n} + \boldsymbol{e}_2, \\ \lambda_{\text{hv}} & \text{if } N_3 \le n_{\text{d}} + n_{\text{v}} \le N - 1,\ \boldsymbol{n}' = \boldsymbol{n} + \boldsymbol{e}_2, \\ n_{\text{d}}\mu_{\text{d}} & \text{if } \boldsymbol{n}' = \boldsymbol{n} - \boldsymbol{e}_1, \\ n_{\text{v}}\mu_{\text{v}} & \text{if } \boldsymbol{n}' = \boldsymbol{n} - \boldsymbol{e}_2, \\ 0 & \text{in other cases,} \end{cases} \tag{1.2}$$

where $\lambda_{\text{d}} := \lambda_{\text{od}} + \lambda_{\text{hd}}$, $\lambda_{\text{v}} := \lambda_{\text{ov}} + \lambda_{\text{hv}}$, $\boldsymbol{e}_1 = (1,0)$, $\boldsymbol{e}_2 = (0,1)$.

The given 2-D MC is strictly continuous with respect to the second component while it is weakly continuous with respect to the first one (for more strong definitions see Appendix). The state diagram of the model and the system of global balance equations (SGBE) for the steady-state probabilities $p(\boldsymbol{n})$, $\boldsymbol{n} \in S$ are shown in [27]. The existence of the stationary regime is proved by the fact that all states of finite-dimensional state space S are communicating.

Desired QoS metrics are determined via the stationary distribution of the initial model. Let P_x denote the blocking probability of the x-calls, $x \in \{\text{hv,ov,hd,od}\}$. Then by using the PASTA theorem [37] we obtain:

$$P_{\text{hv}} := \sum_{\boldsymbol{n} \in S} p(\boldsymbol{n}) \delta(n_{\text{d}} + n_{\text{v}}, N), \tag{1.3}$$

$$P_{\text{ov}} := \sum_{\boldsymbol{n} \in S} p(\boldsymbol{n}) I(n_{\text{d}} + n_{\text{v}} \ge N_3), \tag{1.4}$$

$$P_{\text{hd}} := \sum_{\boldsymbol{n} \in S} p(\boldsymbol{n}) I(n_{\text{d}} + n_{\text{v}} \ge N_2), \tag{1.5}$$

$$P_{\text{od}} := \sum_{\boldsymbol{n} \in S} p(\boldsymbol{n}) I(n_{\text{d}} + n_{\text{v}} \ge N_1), \tag{1.6}$$

where $I(A)$ denotes the indicator function of event A and $\delta(i,j)$ represents Kronecker's symbols.

The mean number of busy channels $\tilde{N}$ is also calculated via stationary distribution as follows:

$$\tilde{N} := \sum_{k=1}^{N} k p\,(k), \tag{1.7}$$

where $p\,(k) = \sum_{\boldsymbol{n} \in S} p\,(\boldsymbol{n})\,\delta\,(n_{\text{d}} + n_{\text{v}}, k)$, $k = \overline{1, N}$, are marginal probability mass functions.

Stationary distribution is determined as a result of solution of an appropriate SGBE of the given 2-D MC. However, to solve the last problem one requires laborious computation efforts for large values of N since the corresponding SGBE has no explicit solution. Very often the solution of such problems is evident if the corresponding 2-D MC has a reversibility property [16] and hence for it there exists a stationary distribution of multiplicative form. However, SGBE has a multiplicative solution only in a special case when $N_1 = N_2 = N_3 = N$ (even in this case there are known computational difficulties). However, by applying Kolmogorov criteria [16] it is easily verified that the given 2-D MC is not reversible. Indeed, according to the mentioned criteria the necessary reversibility condition of 2-D MC consists in the fact that if there exists the transition from state (i,j) to state (i',j'), then there must also be the reverse transition from state (i',j') to state (i,j). However, for the considered MC this condition is not fulfilled. So by the relations (1.2) in the given MC there exists the transition $(n_d,n_v)\rightarrow(n_d - 1,n_v)$ with intensity $n_d\mu_d$ where $n_d+n_v\geq N_2$, but the inverse transition does not exist.

In [27] a recursive technique was proposed for solution of the above-mentioned SGBE. It requires a multiple inversion calculation of certain matrices of sufficiently large dimensions that in itself is a complex calculating procedure. To overcome the mentioned difficulties, a new efficient and refined approximate method for calculation of stationary distribution of the given model is suggested below (see Appendix). The proposed method, because of the correct selection of state space splitting of the corresponding 2-D MC allows one to reduce the solution of the problem considered to calculation by explicit formulae which contain the known (even tabulated) stationary distributions of classical queuing models.

For correct application of phase-merging algorithms (PMA) it is assumed below that $\lambda_v >> \lambda_d$ and $\mu_v >> \mu_d$. This assumption is not extraordinary for an integrated voice/data CWN, since this is a regime that commonly occurs in multimedia networks, in which wide-band d-calls have both longer holding times and significantly smaller arrival rates than narrow-band v-calls, for example see [5, 11]. Moreover, it is more important to note, that the final results shown below are independent of traffic parameters, and are determined from their ratio, i.e. the developed approach can provide a refined approximation even when parameters of heterogeneous traffic are only moderately distinctive.

The following splitting of state space (1.1) is examined:

$$S = \bigcup_{k=0}^{N_2} S_k, \quad S_k \bigcap S_{k'} = \varnothing, k \neq k', \tag{1.8}$$

where $S_k := \{\boldsymbol{n} \in S : n_d = k\}$.

Note 1.1. The assumption above meets the major requirement for correct use of PMA [17]: the state space of the initial model must split into classes, such that transition probabilities within classes are essentially higher than those between states of different classes. Indeed, it is seen from (1.2) that the above-mentioned requirement is fulfilled when using splitting (1.8).

Furthermore, state classes S_k combine into separate merged states $<k>$ and the following merging function in state space S is introduced:

$$U(\boldsymbol{n}) = < k > \text{ if } \boldsymbol{n} \in S_k,\ k = \overline{0, N_2}\ . \tag{1.9}$$

Function (1.9) determines a merged model which is a 1-D MC with the state space $\tilde{S} := \left\{ < k >: k = \overline{0, N_2} \right\}$. Then, according to PMA, the stationary distribution of the initial model approximately equals:

$$p(k,i) \approx \rho_k(i)\pi(< k >),\ (k,i) \in S_k,\ k = \overline{0, N_2}\ , \tag{1.10}$$

where $\{\rho_k(i) : (k,i) \in S_k\}$ is the stationary distribution of a split model with state space S_k and $\left\{ \pi\ (< k >) :< k >\in \tilde{S} \right\}$ is the stationary distribution of the merged model, respectively.

The state diagram of the split model with state space S_k is shown in Fig. 1.1a. By using (1.2) we conclude that the elements of the generating matrix of this 1-D birth-and-death process (BDP) $q_k(i,j)$ are obtained as follows:

$$q_k\ (i,j) = \begin{cases} \lambda_{\text{v}} & \text{if } i \le N_3 - k - 1\,,\ j = i + 1, \\ \lambda_{\text{hv}} & \text{if } N_3 - k < i \le N,\ j = i + 1, \\ i\mu_{\text{v}} & \text{if } j = i - 1, \\ 0 & \text{in other cases.} \end{cases}$$

So, stationary distribution within class S_k is the same as the $M|M|N\text{–}k|N\text{–}k$ queuing system where service rate of each channel is constant, μ_{v} and arrival rates are variable quantities

$$\begin{cases} \lambda_{\text{v}} & \text{if } i < N_3 - k, \\ \lambda_{\text{hv}} & \text{if } j \ge N_3 - k. \end{cases}$$

(a)

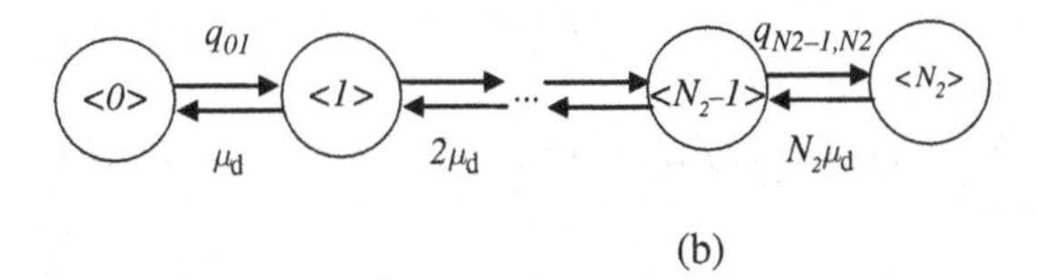

(b)

Fig. 1.1 State diagram of a split model with state space S_k, $k = 0,1,\ldots,N_2$ (**a**) and merged model (**b**)

Hence the desired stationary distribution is

$$\rho_k(i) = \begin{cases} \frac{\nu_v^i}{i!}\rho_k(0) & \text{if } 1 \le i \le N_3 - k, \\ \left(\frac{\nu_v}{\nu_{hv}}\right)^{N_3-k} \frac{\nu_{hv}^i}{i!}\rho_k(0) & \text{if } N_3 - k + 1 \le i \le N - k, \end{cases} \tag{1.11}$$

where $\rho_k(0) = \left(\sum_{i=0}^{N_3-k} \frac{\nu_v^i}{i!} + \left(\frac{\nu_v}{\nu_{hv}}\right)^{N_3-k} \sum_{i=N_3-k+1}^{N-k} \frac{\nu_{hv}^i}{i!} \right)^{-1}$, $\nu_v := \lambda_v/\mu_v$, $\nu_{hv} := \lambda_{hv}/\mu_v$.

Then, from (1.2) to (1.11) by means of PMA elements of the generating matrix of a merged model $q\left(< k >, < k' >\right)$, $< k >, < k' > \in \tilde{S}$ are found:

$$q\left(< k >, < k' >\right) = \begin{cases} \lambda_d \sum_{i=0}^{N_1-k-1} \rho_k(i) + \lambda_{hd} \sum_{i=N_1-1}^{N_2-k-1} \rho_k(i) & \text{if } 0 \le k \le N_1 - 1,\ k' = k+1, \\ \lambda_{hd} \sum_{i=0}^{N_2-k-1} \rho_k(i) & \text{if } N_1 \le k \le N_2 - 1,\ k' = k+1, \\ k\mu_d & \text{if } k' = k-1, \\ 0 & \text{in other cases.} \end{cases} \tag{1.12}$$

The latter formula allows determination of the stationary distribution of a merged model. It coincides with an appropriate distribution of state probabilities of a 1-D BDP, for which transition intensities are determined in accordance with (1.12). Consequently, the stationary distribution of a merged model is determined as (see Fig. 1.1b):

$$\pi(< k >) = \frac{\pi\left(< 0 >\right)}{k!\mu_d^k} \prod_{i=1}^{k} q\left(< k-1 >, < k >\right),\ k = \overline{1, N_2}, \tag{1.13}$$

where $\pi\left(< 0 >\right) = \left(1 + \sum_{k=1}^{N_2} \frac{1}{k!\mu_d^k} \prod_{i=1}^{k} q\left(< k-1 >, < k >\right)\right)^{-1}$.

Then by using (1.11) and (1.13) from (1.10) the stationary distribution of the initial 2-D MC can be found. So, summarizing the above and omitting the complex algebraic transformations the following approximate formulae for calculation of QoS metrics (1.3), (1.4), (1.5), (1.6), and (1.7) can be suggested:

$$P_{hv} \approx \sum_{k=0}^{N_2} \pi\left(< k >\right) \rho_k\left(N - k\right); \tag{1.14}$$

$$P_{ov} \approx \sum_{k=0}^{N_2} \pi\left(< k >\right) \sum_{i=N_3-k}^{N-k} \rho_k(i); \tag{1.15}$$

$$P_{\text{hd}} \approx \sum_{k=0}^{N_2} \pi\,(<k>) \sum_{i=N_2-k}^{N-k} \rho_k(i)\,; \tag{1.16}$$

$$P_{\text{od}} \approx \sum_{k=0}^{N_1-1} \pi\,(<k>) \sum_{i=N_1-k}^{N-k} \rho_k(i) + \sum_{k=N_1}^{N_2} \pi\,(<k>)\,; \tag{1.17}$$

$$\tilde{N} \approx \sum_{i=1}^{N} i \sum_{k=0}^{f_{N_2}(i)} \pi\,(<k>)\rho_k\,(i-k). \tag{1.18}$$

$$\text{Hereinafter } f_k\,(x) = \begin{cases} x & \text{if } 1 \le x \le k, \\ k & \text{if } k \le i \le N. \end{cases}$$

Now we can develop the following algorithm to calculate the QoS metrics of the investigated multi-parametric CAC for a similar model with wide-band d-calls, i.e. when $b>1$.

Step 1. For $k = 0,1,\ldots,[N_2/b]$ calculate the following quantities

$$\rho_k(i) = \begin{cases} \frac{\nu_v^i}{i!}\rho_k\,(0) & \text{if } 1 \le i \le N_3 - kb, \\ \left(\frac{\nu_v}{\nu_{hv}}\right)^{N_3-kb} \frac{\nu_{hv}^i}{i!}\rho_k\,(0) & \text{if } N_3 - kb + 1 \le i \le N - kb, \end{cases}$$

$$\text{where } \rho_k(0) = \left(\sum_{i=0}^{N_3-kb} \frac{\nu_v^i}{i!} + \left(\frac{\nu_v}{\nu_{hv}}\right)^{N_3-kb} \sum_{i=N_3-kb+1}^{N-kb} \frac{\nu_{hv}^i}{i!} \right)^{-1};$$

$$\pi(<k>) = \frac{\pi\,(<0>)}{k!\mu_d^k} \prod_{i=1}^{k} q(<k-1>,<k>),$$

$$\text{where } \pi(<0>) = \left(1 + \sum_{k=1}^{[N_2/b]} \frac{1}{k!\mu_d^k} \prod_{i=1}^{k} q(<k-1>,<k>) \right)^{-1},$$

$$q(<k>,\ <k'>) = \begin{cases} \lambda_d \sum\limits_{i=0}^{N_1-kb-1} \rho_k\,(i) + \lambda_{hd} \sum\limits_{i=N_1-kb}^{N_2-kb-1} \rho_k\,(i) & \text{if } 0 < k < [N_1/b] - 1,\ k' = k+1, \\ \lambda_{hd} \sum\limits_{i=0}^{N_2-kb-1} \rho_k\,(i) & \text{if } [N_1/b] < k < [N_2/b] - 1,\ k' = k+1, \\ k\mu_d & \text{if } k' = k-1, \\ 0 & \text{in other cases.} \end{cases}$$

Step 2. Calculate the approximate values (AV) of QoS metrics:

$$P_{\text{hv}} \approx \sum_{k=0}^{[N_2/b]} \pi\,(<k>)\,\rho_k\,(N-kb)\,;$$

$$P_{\text{ov}} \approx \sum_{k=0}^{[N_2/b]} \pi\,(<k>) \sum_{i=N_3-kb}^{N-kb} \rho_k\,(i)\,;$$

$$P_{\text{hd}} \approx \sum_{k=0}^{[N_2/b]} \pi\,(<k>) \sum_{i=N_2-kb}^{N-kb} \rho_k(i)\,;$$

$$P_{\text{od}} \approx \sum_{k=0}^{[N_1/b]-1} \pi\,(<k>) \sum_{i=N_1-kb}^{N-kb} \rho_k(i) + \sum_{k=[N_1/b]}^{[N_2/b]} \pi\,(<k>)\,;$$

$$\tilde{N} \approx \sum_{i=1}^{N} i \sum_{k=0}^{f_{[N_2/b]}(i)} \pi\,(<k>)\,\rho_k\,(i-k).$$

Henceforth $[x]$ denotes the integer part of x.

Now consider some important special cases of the investigated multi-parametric CAC (for the sake of simplicity consider case $b=1$).

1. CAC based on Complete Sharing (CS). Under the given CAC strategy, no distinction is made between voice calls and data calls for channel access, i.e. it is assumed that $N_1=N_2=N_3=N$. In other words, we have the 2-D Erlang's loss model. It is obvious, that in this case the blocking probabilities of calls from heterogeneous traffic are equal to each other, i.e. this probability according to the PASTA theorem coincides with the probability that a call of any type on arrival finds all channels of a cell occupied. Then from (1.11) to (1.18) particularly we get the following convolution algorithms for calculation of QoS metrics in the given model:

$$P_{\text{hv}} = P_{\text{ov}} = P_{\text{hd}} = P_{\text{od}} \approx \sum_{k=0}^{N} E_{\text{B}}\,(\nu_{\text{v}}, N-k)\,\pi\,(<k>), \tag{1.19}$$

$$\tilde{N} \approx \sum_{i=1}^{N} i \sum_{k=0}^{i} \theta_{i-k}\,(\nu_{\text{v}}, N-k)\,\pi\,(<k>). \tag{1.20}$$

Here

$$\pi\,(<k>) = \frac{\nu_{\text{d}}^k}{k!} \prod_{i=0}^{k-1} (1 - E_{\text{B}}\,(\nu_{\text{v}}, N-i))\,\pi\,(<0>)\;,\; k = \overline{1,N}, \tag{1.21}$$

where $\pi\,(<0>) = \left(1 + \sum_{k=1}^{N} \frac{\nu_{\text{d}}^k}{k!} \prod_{i=0}^{k-1} (1 - E_{\text{B}}\,(\nu_{\text{v}}, N-i))\right)^{-1}.$

Henceforth $E_{\text{B}}(\nu, m)$ denotes Erlang's B-formula for the model *M/M/m/m* with load ν Erl, and $\theta_i(\nu, m)$, $i=0,1,\ldots,m$, denotes the steady-state probabilities in the same model, i.e.

$$\theta_i\,(\nu, m) = \left(\frac{\nu_{\text{v}}^i}{i!}\right) \left(\sum_{j=0}^{m} \frac{\nu_{\text{v}}^j}{j!}\right)^{-1}, \quad i = \overline{0,m}\;;\; E_{\text{B}}\,(\nu, m) := \theta_m\,(\nu, m)\,. \tag{1.22}$$

Note that the above-developed analytic results for the CS-strategy are similar in spirit to the algorithm for nearly decomposable 2-D MC proposed in [11].

2. CAC with a Single Parameter. Given the strategy tell the difference between voice calls and data calls but do not take into account distinctions between original and handover calls within each traffic, i.e. it is assumed that $N_1 = N_2$ and $N_3 = N$ where $N_2 < N_3$. In other words, there is only one threshold parameter. For this case from (1.11) to (1.18) we get the following approximate formulae for calculating the blocking probabilities of v-calls (P_{v}) and d-calls (P_{d}) and mean number of busy channels:

$$P_{\text{v}} = P_{\text{hv}} = P_{\text{ov}} \approx \sum_{k=0}^{N_2} E_{\text{B}}(\nu_{\text{v}}, N-k)\pi\,(<k>), \tag{1.23}$$

$$P_{\text{d}} = P_{\text{hd}} = P_{\text{od}} \approx \sum_{k=0}^{N_2} \pi\,(<k>) \sum_{i=N_2-k}^{N-k} \theta_i\,(\nu_{\text{v}}, N-k)\,, \tag{1.24}$$

$$\tilde{N} \approx \sum_{i=1}^{N} i \sum_{k=0}^{f_{N_2}(i)} \theta_{i-k}\,(\nu_{\text{v}}, N-k)\,\pi\,(<k>)\,. \tag{1.25}$$

Here

$$\pi(<k>) = \frac{\nu_{\text{d}}^k}{k!}\prod_{i=1}^{k}\Lambda(i)\pi(<0>)\,,\; k = \overline{1, N_2}, \tag{1.26}$$

where $\pi(<0>) = \left(1 + \sum_{j=1}^{N_2} \frac{\nu_{\text{d}}^j}{j!}\prod_{i=1}^{j}\Lambda(i)\right)^{-1}$, $\Lambda(i) := \theta_0(\nu_{\text{v}}, N-i+1)\sum_{j=0}^{N_2-i}\frac{\nu_{\text{v}}^j}{j!}$.

3. A mono-service CWN with guard channels. The last results can be interpreted for the model of an isolated cell in a mono-service CWN with guard channels for h-calls, i.e. for a model in which distinctions between original and handover calls of single traffic are taken into account. A brief description of the model is as follows. The network supports only original and handover calls of single traffic that arrive according to Poisson processes with rates λ_{o} and λ_{h}, respectively. Assume that the o-call (h-call) holding times have an exponential distribution with mean μ_{o} (μ_{h}) but their parameters are different, i.e. generally speaking $\mu_{\text{o}} \neq \mu_{\text{h}}$, for example see [38], pages 267–268.

In the cell the mentioned one-parametric CAC strategy based on the guard channels scheme is realized in the following way [12]. If upon arrival of an h-call, there is at least one free channel, this call seizes one of the free channels; otherwise the h-call is dropped. A newly arrived o-call is accepted only in the case where there are at least $g+1$ free channels (i.e. at most N–g–1 busy channels), otherwise the o-call is blocked. Here $g \geq 0$ denotes the number of guard channels that are reserved only for h-calls.

By using the above-described approach and omitting the known intermediate transformations we conclude that QoS metrics of the given model are calculated as follows:

$$P_{\mathrm{o}} \approx \sum_{k=0}^{N-g} \pi\,(< k >) \sum_{i=N-g-k}^{N-k} \theta_i(\nu_{\mathrm{h}}, N-k), \tag{1.27}$$

$$P_{\mathrm{h}} \approx \sum_{k=0}^{N-g} E_{\mathrm{B}}(\nu_{\mathrm{h}}, N-k)\pi(\leq k \geq), \tag{1.28}$$

$$\tilde{N} \approx \sum_{i=1}^{N} i \sum_{k=0}^{f_{N-g}(i)} \theta_{i-k}\,(\nu_h, N-k)\,\pi(< k >)\,. \tag{1.29}$$

Here

$$\pi(< k >) = \frac{\nu_{\mathrm{o}}^k}{k!} \prod_{i=1}^{k} \Lambda(i)\pi(< 0 >)\,,\ k = \overline{1, N-g}, \tag{1.30}$$

where

$$\nu_{\mathrm{o}} = \lambda_{\mathrm{o}}/\mu_{\mathrm{o}},\, \nu_{\mathrm{h}} = \lambda_{\mathrm{h}}/\mu_{\mathrm{h}};\, \pi(< 0 >) = \left(1 + \sum_{j=1}^{N-g} \frac{\nu_{\mathrm{o}}^j}{j!} \prod_{i=1}^{j} \Lambda(i)\right)^{-1},$$

$$\Lambda(i) = \theta_0(\nu_{\mathrm{h}}, N-i+1) \sum_{j=0}^{N-g-i} \frac{\nu_{\mathrm{h}}^j}{j!}\,.$$

Formulae (1.27), (1.28), (1.29), and (1.30) coincide with ones for CAC with a single parameter in integrated voice/data networks if we set $g{:=}N{-}N_2$, $\nu_{\mathrm{o}}{:=}\nu_{\mathrm{d}}$, $\nu_{\mathrm{h}}{:=}\nu_{\mathrm{v}}$. And in the case where $g=0$ we get the results for CAC based on the CS-strategy, see (1.19), (1.20), and (1.21). Also from (1.28) we get the following un-improvable limits for P_{h} which will be very useful for solution of optimization problems in the future:

$$E_{\mathrm{B}}\,(\nu_{\mathrm{h}}, N) \leq P_{\mathrm{h}} \leq E_{\mathrm{B}}\,(\nu_{\mathrm{h}}, g)\,. \tag{1.31}$$

In the proposed algorithms the computational procedures contain the well-known Erlang's B-formula as well as expressions within that formula which have even been tabulated [9, 34]. Thus, the complexity of the proposed algorithms to calculate QoS metrics of the investigated multi-parametric CAC based on guard channels are almost congruous to that of Erlang's B-formula. Direct calculations by Erlang's B-formula bring known difficulties at large values of N because of large factorials and exponents. To overcome these difficulties the known effective recurrent formulae can be used, for example see [12].

1.2 CAC Based on a Cutoff Strategy

Here we describe an alternative CAC in integrated voice/data networks which is based on a cutoff strategy. A more detailed description of the given CAC is as follows.

As in CAC based on guard channels, we assume that on arrival an hv-call is accepted as long as at least one free channel is available; otherwise it is blocked. With the purpose of definition of CAC based on a cutoff strategy for calls of other types three parameters R_1, R_2, and R_3 where $1 \le R_1 \le R_2 \le R_3 \le N$ are introduced. Then the proposed CAC defines the following rules for admission of heterogeneous calls: an od-call (respectively, hd-call and ov-call) is accepted only if the number of calls of the given type in progress is less than R_1 (respectively, R_2 and R_3) and a free channel is available; otherwise it is blocked.

For the sake of simplicity we shall assume that $b=1$. The case where $b>1$ is straightforward (see Sect. 1.1). The state of the system under the given CAC at any time is also described by the 2-D vector $\boldsymbol{n}=(n_\mathrm{d}, n_\mathrm{v})$, where n_d (respectively, n_v) is the number of data (respectively, voice) calls in the channels. Then the state space of the appropriate 2-D MC is given by:

$$S := \left\{\boldsymbol{n} : n_\mathrm{d} = \overline{0, R_2},\ n_\mathrm{v} = \overline{0, N};\ n_\mathrm{d} + n_\mathrm{v} \le N\right\}. \tag{1.32}$$

Note 1.2. Hereinafter, for simplicity, we use the same notations for state spaces, stationary distribution etc. in different CAC strategies. This should not lead to misunderstanding, as it will be clear what model is considered from the context.

The elements of the generating matrix of the appropriate 2-D MC in this case is determined as follows:

$$q\left(\boldsymbol{n}, \boldsymbol{n}'\right) = \begin{cases} \lambda_\mathrm{d} & \text{if } n_\mathrm{d} \le R_1 - 1,\ \boldsymbol{n}' = \boldsymbol{n} + \boldsymbol{e}_1, \\ \lambda_\mathrm{hd} & \text{if } R_1 \le n_\mathrm{d} \le R_2 - 1,\ \boldsymbol{n}' = \boldsymbol{n} + \boldsymbol{e}_1, \\ \lambda_\mathrm{v} & \text{if } n_\mathrm{v} \le R_3 - 1,\ \boldsymbol{n}' = \boldsymbol{n} + \boldsymbol{e}_2, \\ \lambda_\mathrm{hv} & \text{if } R_3 \le n_\mathrm{v} \le N - 1,\ \boldsymbol{n}' = \boldsymbol{n} + \boldsymbol{e}_2, \\ n_\mathrm{d}\mu_\mathrm{d} & \text{if } \boldsymbol{n}' = \boldsymbol{n} - \boldsymbol{e}_1, \\ n_\mathrm{v}\mu_\mathrm{v} & \text{if } \boldsymbol{n}' = \boldsymbol{n} - \boldsymbol{e}_2, \\ 0 & \text{in other cases.} \end{cases} \tag{1.33}$$

The blocking probability of hv-calls and the mean number of busy channels are defined similarly to (1.3) and (1.7), respectively. The other QoS metrics are defined as the following marginal distributions of the initial chain:

$$P_\mathrm{ov} := \sum_{\boldsymbol{n} \in S} p(\boldsymbol{n}) I(n_\mathrm{v} \ge R_3), \tag{1.34}$$

$$P_\mathrm{hd} := \sum_{\boldsymbol{n} \in S} p(\boldsymbol{n}) \delta(n_\mathrm{d}, R_2) + \sum_{\boldsymbol{n} \in S} p(\boldsymbol{n}) \delta\left(n_\mathrm{d} + n_\mathrm{v}, N\right) I\left(n_\mathrm{d} \le R_2\right), \tag{1.35}$$

$$P_\mathrm{od} := \sum_{\boldsymbol{n} \in S} p(\boldsymbol{n}) I(n_\mathrm{d} \ge R_1) + \sum_{\boldsymbol{n} \in S} p(\boldsymbol{n}) \delta\left(n_\mathrm{d} + n_\mathrm{v}, N\right) I\left(n_\mathrm{d} \le R_1\right). \tag{1.36}$$

Unlike CAC based on the guard channels strategy, it is easily shown that here there is no circulation flow in the state diagram of the underlying 2-D MC, i.e. it is reversible [16]. In other words, there is a general solution of the system of local balance equations (SLBE) in this chain. Therefore, we can express any state probability $p(n_d,n_v)$ via state probability $p(0,0)$ by choosing any path between these states in the state diagram. So, in the case where $R_2 + R_3 \le N$ we get the following multiplicative solution for the stationary distribution of the underlying 2-D MC:

$$p(n_d,n_v)=\begin{cases}\frac{\nu_d^{n_d}}{n_d!}\cdot\frac{\nu_v^{n_v}}{n_v!}\cdot p(0,0), & \text{if } n_d\le R_1,\ n_v\le R_3,\\ \frac{\nu_d^{n_d}}{n_d!}\cdot\frac{\nu_{hv}^{n_v}}{n_v!}\cdot\left(\frac{\nu_v}{\nu_{hv}}\right)^{R_3}\cdot p(0,0), & \text{if } n_d\le R_1,\ R_3<n_v\le N,\\ \frac{\nu_{hd}^{n_d}}{n_d!}\cdot\frac{\nu_v^{n_v}}{n_v!}\cdot\left(\frac{\nu_d}{\nu_{hd}}\right)^{R_1}\cdot p(0,0), & \text{if } R_1<n_d\le R_2,\ n_v\le R_3,\\ \frac{\nu_{hd}^{n_d}}{n_d!}\cdot\frac{\nu_{hv}^{n_v}}{n_v!}\cdot\left(\frac{\nu_d}{\nu_{hd}}\right)^{R_1}\cdot\left(\frac{\nu_v}{\nu_{hv}}\right)^{R_3}\cdot p(0,0), & \text{if } R_1<n_d\le R_2,\ R_3<n_v\le N,\end{cases} \tag{1.37}$$

where $p(0,0)$ is determined from the normalizing condition:

$$p(0,0)=\left(\sum_{n\in S_1}\frac{\nu_d^{n_d}}{n_d!}\cdot\frac{\nu_v^{n_v}}{n_v!}+\left(\frac{\nu_v}{\nu_{hv}}\right)^{R_3}\sum_{n\in S_2}\frac{\nu_d^{n_d}}{n_d!}\cdot\frac{\nu_{hv}^{n_v}}{n_v!}+\left(\frac{\nu_d}{\nu_{hd}}\right)^{R_1}\right.$$
$$\left.\sum_{n\in S_3}\frac{\nu_{hd}^{n_d}}{n_d!}\cdot\frac{\nu_v^{n_v}}{n_v!}+\left(\frac{\nu_d}{\nu_{hd}}\right)^{R_3}\left(\frac{\nu_v}{\nu_{hv}}\right)^{R_3}\sum_{n\in S_4}\frac{\nu_{hd}^{n_d}}{n_d!}\cdot\frac{\nu_{hv}^{n_v}}{n_v!}\right)^{-1}$$

Here we use the following notations: $\nu_d := \lambda_d/\mu_d$, $\nu_{hd} := \lambda_{hd}/\mu_d$;

$$S_1 := \{\mathbf{n}\in S : n_d\le R_1, n_v\le R_3\},\ S_2 := \{\mathbf{n}\in S : n_d\le R_1, R_3+1\le n_v\le N\},$$
$$S_3 := \{\mathbf{n}\in S : R_1+1\le n_d\le R_2, n_v\le R_3\},$$
$$S_4 := \{\mathbf{n}\in S : R_1+1\le n_d\le R_2, R_3+1\le n_v\le N\}.$$

In the case where $R_2 + R_3 > N$ the stationary distribution has the following form:

$$p(n_d,n_v)=\begin{cases}\frac{\nu_d^{n_d}}{n_d!}\cdot\frac{\nu_v^{n_v}}{n_v!}\cdot p(0,0), & \text{if } 0\le n_d\le R_1,\ 0\le n_v\le R_3,\\ \frac{\nu_{hd}^{n_d}}{n_d!}\cdot\frac{\nu_v^{n_v}}{n_v!}\cdot\left(\frac{\nu_d}{\nu_{hd}}\right)^{R_2}\cdot p(0,0), & \text{if } R_1+1\le n_d\le R_2,\ 0\le n_v\le N-n_d,\\ \frac{\nu_d^{n_d}}{n_d!}\cdot\frac{\nu_{hv}^{n_v}}{n_v!}\cdot\left(\frac{\nu_v}{\nu_{hv}}\right)^{R_3}\cdot p(0,0), & \text{if } 0\le n_d\le N-R_3-1,\ R_3+1\le n_v\le N,\end{cases} \tag{1.38}$$

where

$$p(0,0)=\left(\sum_{n\in T_1}\frac{\nu_{\mathrm{d}}^{n_{\mathrm{d}}}}{n_{\mathrm{d}}!}\cdot\frac{\nu_{\mathrm{v}}^{n_{\mathrm{v}}}}{n_{\mathrm{v}}!}+\left(\frac{\nu_{\mathrm{d}}}{\nu_{\mathrm{hd}}}\right)^{R_1}\sum_{n\in T_2}\frac{\nu_{\mathrm{hd}}^{n_{\mathrm{d}}}}{n_{\mathrm{d}}!}\cdot\frac{\nu_{\mathrm{v}}^{n_{\mathrm{v}}}}{n_{\mathrm{v}}!}+\left(\frac{\nu_{\mathrm{v}}}{\nu_{\mathrm{hv}}}\right)^{R_3}\sum_{n\in T_3}\frac{\nu_{\mathrm{d}}^{n_{\mathrm{d}}}}{n_{\mathrm{d}}!}\cdot\frac{\nu_{\mathrm{hv}}^{n_{\mathrm{v}}}}{n_{\mathrm{v}}}\right)^{-1};$$

$$T_1 := \{\boldsymbol{n}\in S: 0\le n_{\mathrm{d}}\le R_1,\ 0\le n_{\mathrm{v}}\le R_3\},$$
$$T_2 := \{\boldsymbol{n}\in S: R_1+1\le n_{\mathrm{d}}\le R_2, 0\le n_{\mathrm{v}}\le N-n_{\mathrm{d}}\},$$
$$T_3 := \{\boldsymbol{n}\in S: 0\le n_{\mathrm{d}}\le N-R_3-1,\ R_3+1\le n_{\mathrm{v}}\le N\}.$$

The exact method to determine the steady-state probabilities in terms of a multiplicative representation (1.37) [or (1.38)] for large values of N encounters numerical problems such as imprecision and overflow. These are related to the fact that with such a method the entire state space has to be generated, and large factorials and powers close to the zero of the quantities (for low loads) or large values (for high loads) have to be calculated, i.e. there arises the problem of exponent overflow or underflow. Hence we can use the developed approximate method to determine the QoS metrics of the model when using the proposed CAC based on a cutoff strategy even when the state space (1.32) is large.

As in Sect. 1.1, we assume that $\lambda_{\mathrm{v}}>>\lambda_{\mathrm{d}}$ and $\mu_{\mathrm{v}}>>\mu_{\mathrm{d}}$ and examine the following splitting of the state space (1.32):

$$S=\bigcup_{k=0}^{R_2}S_k,\quad S_k\bigcap S_{k'}=\varnothing,\ k\ne k',$$

where

$$S_k := \{\boldsymbol{n}\in S: n_{\mathrm{d}}=k\}.$$

Next classes of states S_k are combined into individual merged states $<k>$ and in (1.32) the merged function with range $\tilde{S} := \{<k>: k=0,1,\ldots,R_2\}$ which is similar to (1.9) is introduced. As in the exact algorithm in order to find the stationary distribution within splitting classes S_k we will distinguish two cases: (1) $R_2+R_3\le N$ and (2) $R_2+R_3>N$.

In the first case the elements of the generating matrix of the appropriate 1-D BDP are the same for all splitting models, i.e.

$$q_k(i,j)=\begin{cases}\lambda_{\mathrm{v}} & \text{if } i\le R_3-1,\ j=i+1,\\ \lambda_{\mathrm{hv}} & \text{if } R_3\le i\le N-1,\ j=i+1,\\ i\mu_{\mathrm{v}} & \text{if } j=i-1,\\ 0 & \text{in other cases.}\end{cases}$$

From the last formula we conclude that the stationary distribution within class S_k is the same as the $M|M|N–k|N–k$ queuing system with state-dependent arrival rates and constant service rate of each channel, i.e.

$$\rho_k(i) = \begin{cases} \frac{\nu_{\mathrm{v}}^i}{i!}\rho_k(0) & \text{if } 1 \le i \le R_3, \\ \left(\frac{\nu_{\mathrm{v}}}{\nu_{\mathrm{hv}}}\right)^{R_3} \frac{\nu_{\mathrm{hv}}^i}{i!}\rho_k(0) & \text{if } R_3+1 \le i \le N-k, \end{cases} \tag{1.39}$$

where

$$\rho_k(0) = \left(\sum_{i=0}^{R_3} \frac{\nu_{\mathrm{v}}^i}{i!} + \left(\frac{\nu_{\mathrm{v}}}{\nu_{\mathrm{hv}}}\right)^{R_3} \sum_{i=R_3+1}^{N-k} \frac{\nu_{\mathrm{hv}}^i}{i!} \right)^{-1}.$$

So, from (1.33) to (1.39) we conclude that the elements of the generating matrix of the merged model are

$$q(<k>, <k'>) = \begin{cases} \lambda_{\mathrm{d}}\left(1-\rho_k(N-k)\right) & \text{if } 0 \le k \le R_1-1,\ k'=k+1, \\ \lambda_{\mathrm{hd}}\left(1-\rho_k(N-k)\right) & \text{if } R_1 \le k \le R_2-1,\ k'=k+1, \\ k\mu_{\mathrm{d}} & \text{if } k'=k-1, \\ 0 & \text{in other cases.} \end{cases} \tag{1.40}$$

Distribution of the merged model is calculated by using (1.40) and has the following form:

$$\pi(<k>) = \frac{\pi(<0>)}{k!\mu_{\mathrm{d}}^k} \prod_{i=1}^{k} q(<k-1>, <k>),\ k=\overline{1,R_2}, \tag{1.41}$$

where

$$\pi(<0>) = \left(1 + \sum_{k=1}^{R_2} \frac{1}{k!\mu_{\mathrm{d}}^k} \prod_{i=1}^{k} q(<k-1>, <k>)\right)^{-1}.$$

Finally, the following approximate formulae to calculate the desired QoS metrics when using the proposed CAC based on a cutoff strategy are obtained:

$$P_{\mathrm{hv}} \approx \sum_{k=0}^{R_2} \pi(<k>)\,\rho_k(N-k); \tag{1.42}$$

$$P_{\mathrm{ov}} \approx \sum_{k=0}^{R_2} \pi(<k>) \sum_{i=R_3}^{N-k} \rho_k(i); \tag{1.43}$$

$$P_{\mathrm{hd}} \approx \pi(<R_2>) + \sum_{k=0}^{R_2-1} \pi(<k>)\rho_k(N-k); \tag{1.44}$$

$$P_{\mathrm{od}} \approx \sum_{k=R_1}^{R_2} \pi(<k>) + \sum_{k=0}^{R_1-1} \pi(<k>)\rho_k(N-k); \tag{1.45}$$

$$N_{\mathrm{av}} \approx \sum_{k=1}^{N} k \sum_{i=0}^{f_{R_2}(k)} \pi(<i>)\,\rho_i(k-i). \tag{1.46}$$

In the second case (i.e. when $R_2+R_3>N$) the distributions for splitting models with state space S_k for $k = 0,1, \ldots ,N-R_3-1$ are calculated using relations (1.39), while distributions for splitting models with state space S_k for $k=N-R_3,\ldots,R_2$ coincide with distributions of the model $M/M/N-k/N-k$ with loadν_v Erl, see (1.22). And all stages of the developed procedure to calculate the QoS metrics are the same as for the first case except for the calculation of P_{ov}. The latter QoS metric in this case is calculated as follows:

$$P_{ov} \approx \sum_{k=0}^{N-R_3} \pi\left(<k>\right) \sum_{i=R_3}^{N-k} \rho_k\left(i\right) + \sum_{k=N=R_3+1}^{R_2} \pi\left(<k>\right) \rho_k\left(N-k\right). \quad (1.47)$$

Now consider some special cases. First of all note that CAC based on the CS-strategy is a special case of the proposed one when $R_1 = R_2 = R_3 = N$. It is important to note that if we set in the developed approximate algorithm the indicated value of parameters we obtain exactly the results which were established in Sect. 1.1, see (1.19), (1.20), and (1.21).

1. CAC with a Single Parameter. As in Sect. 1.1, let us examine a subclass of the investigated CAC in which distinction is made only between voice and data traffic, i.e. it is assumed that $R_1 = R_2$ and $R_3 = N$ where $R_2<R_3$. For this case from (1.39) to (1.46) we get the following simple approximate formulae for calculating the blocking probabilities of v-calls (P_v) and d-calls (P_d):

$$P_v = P_{hv} = P_{ov} \approx \sum_{k=0}^{R_2} \pi\left(<k>\right) E_B\left(\nu_v, N-k\right), \quad (1.48)$$

$$P_d = P_{hd} = P_{od} \approx \sum_{k=0}^{R_2-1} \pi\left(<k>\right) E_B\left(\nu_v, N-k\right) + \pi\left(<R_2>\right). \quad (1.49)$$

Here

$$\pi\left(<k>\right) = \frac{\nu_d^k}{k!} \prod_{i=0}^{k-1} \left(1 - E_B\left(\nu_d, N-i\right)\right), \; k = \overline{1, R_2}, \quad (1.50)$$

where

$$\pi\left(<0>\right) = \left(1 + \sum_{k=1}^{R_2} \frac{\nu_d^k}{k!} \prod_{i=0}^{k-1} \left(1 - E_B\left(\nu_d, N-i\right)\right)\right)^{-1}.$$

The mean number of busy channels is calculated as follows:

$$\tilde{N} \approx \sum_{i=1}^{N} i \sum_{k=0}^{f_{R_2}(i)} \theta_{i-k}\left(\nu_v, N-k\right) \pi\left(<k>\right). \quad (1.51)$$

Note that if in formulae (1.48), (1.49), (1.50), and (1.51) we set $R_2 = N$ then we obtain the results for CAC based on the CS-strategy, see (1.19), (1.20), and (1.21).

2. Mono-Service CWN with Individual Pools for Heterogeneous Calls. In a given CAC the entire pool of N channels is divided into three pools, an individual pool consisting of r_o channels (for o-calls alone), r_h channels (for h-calls alone), and a common pool consisting of $N{-}r_o{-}r_h$ channels (for o- and h-calls). Assume that $N{>}r_o{+}r_h$, since in the case where $N{=}r_o{+}r_h$ there is a trivial CAC based on a Complete Partitioning (CP) strategy, i.e. the initial system is divided into two separate subsystems where one contains r_h channels for handling only h-calls and the second with r_o channels handles only o-calls.

If there is at least one free channel (either in the appropriate individual or common pool) at the moment an o-call (h-call) arrives, it is accepted for servicing; otherwise the call is lost. Note that the process by which the channels are engaged by heterogeneous calls is realized in the following way. If there is one free channel in its own pool at the moment an o-call (h-call) arrives, it engages a channel from its own individual pool, while if there is no free channel in its own individual pool, the o-call (h-call) utilizes channels from the common pool. Upon completion of servicing of an o-call (h-call) in the individual pool, the relinquished channel is transferred to the common pool if there is an o-call (h-call) present there, while the channel in the common pool that has finished servicing the o-call (h-call) is switched to the appropriate individual pool. This procedure is called the channel reallocation method.

From the above-described model we conclude that it corresponds to a general CAC based on a cutoff strategy in the case where $R_1 = R_2 = N{-}r_h$ and $R_3 = N{-}r_o$. Therefore, taking into account (1.39), (1.40), (1.41), (1.42), (1.43), (1.44), (1.45), and (1.46) we find the following approximate formulae to calculate the QoS metrics of the given model:

$$P_o \approx E_B(\nu_h, N - r_o) \sum_{k=0}^{r_o} \pi(<k>) + \sum_{k=r_o+1}^{N-r_h-1} E_B(\nu_h, N-k)\pi(<k>) + \pi(<N - r_h>), \tag{1.52}$$

$$P_h \approx E_B(\nu_h, N - r_o) \sum_{k=0}^{r_o} \pi(<k>) + \sum_{k=r_o+1}^{N-r_h} E_B(\nu_h, N-k)\pi(<k>), \tag{1.53}$$

$$\tilde{N} \approx \sum_{k=1}^{N-r_h} k \sum_{i=0}^{k} \pi(<i>)\rho_i(k-i) + \sum_{k=N-r_h+1}^{N} k \sum_{i=r_o-N+k}^{N-r_h} \pi(<i>)\rho_i(k-i), \tag{1.54}$$

where

$$\rho_k(i) = \begin{cases} \theta_i(\nu_h, N - r_o), & \text{if } 0 < k \le r_o,\ 0 \le i \le N - r_o, \\ \theta_i(\nu_h, N - k), & \text{if } r_o + 1 \le k \le N - r_h,\ 0 \le i \le N - k; \end{cases} \tag{1.55}$$

$$\pi(<k>) = \begin{cases} \frac{\nu_o^k}{k!}\pi(<0>), & \text{if } 1 \le k \le r_o, \\ \frac{\nu_o^k}{k!} \prod_{i=N-k+1}^{N-r_o} (1 - E_B(\nu_h, i))\pi(<0>), & \text{if } r_o + 1 \le k \le N - r_h, \end{cases} \tag{1.56}$$

$$\pi(<0>) = \left(\sum_{i=0}^{r_o} \frac{\nu_o^k}{k!} + \sum_{k=r_o+1}^{N-r_h} \frac{\nu_o^k}{k!} \prod_{i=N-k+1}^{N-r_h} (1 - E_B(\nu_h, i)) \right)^{-1}.$$

Note that in the special case of $r_o = 0$ the proposed CAC coincides with the one investigated in [10]. It is evident from the derived formulae that in the case of approximate calculation of QoS metrics we don't have to generate the entire state space of the initial model and calculate its stationary distribution in order to calculate the QoS metrics of CAC based on individual pools for heterogeneous calls. These parameters may be found by means of simple computational procedures which contain Erlang's B-formula and terms within that formula. Note that for $r_o = r_h = 0$ this scheme becomes fully accessible by both types of calls, i.e. CAC based on a CS-strategy takes place.

One of the main advantages of the latter CAC in a mono-service CWN can be described as follows. Indeed, in such networks CAC based on the guard channels scheme (see Sect. 1.1) has only one degree of freedom, i.e. if the total number of channels N is fixed then the guard channels scheme has only one controllable parameter g (number of guard channels). However, CAC based on individual pools for heterogeneous calls have two controllable parameters r_o and r_h (size of individual pools), i.e. this CAC has two degrees of freedom. Thus, in the latter case we have additional possibilities to support the QoS parameters at a desired level.

Finally please note that the proposed approximate approach could be applied to studying the model of a cell in which wide-band d-calls have both shorter holding times and significantly larger arrival rates than narrow-band v-calls. Though, as has been noted above, in existing networks such situations arise rarely. This problem is offered to the reader.

1.3 Numerical Results

For realization of the above-derived algorithms a software package was developed to investigate the behavior of QoS metrics as a function of the variation in the values of a cell's load and structure parameters as well as CAC parameters. First we will briefly consider some results for CAC based on the guard channels strategy in an integrated voice/data model with four classes of calls.

The developed approximate formulae allow one – without essential computing difficulties – to carry out an authentic analysis of QoS metrics over any range of change of values of loading parameters of the heterogeneous traffic, satisfying the assumption concerning their ratio (i.e. when $\lambda_v >> \lambda_d$ and $\mu_v >> \mu_d$) and also at any number of channels of cell. Some results are shown in Figs. 1.2, 1.3, and 1.4. The behavior of the studied curves fully confirms all theoretical expectations.

In the given model at a fixed value of the total number of channels (N) it is possible to change the values of three threshold parameters (N_1, N_2, and N_3). In other words, there are three degrees of freedom. Let's note that the increase in value of one of the parameters (in an admissible area) favorably influences the blocking probability of calls of the corresponding type only (see Figs. 1.2 and 1.3). So, in these experiments, the increase in value of parameter N_1 leads to a reduction of blocking probability of od-calls but other blocking probabilities (i.e. P_{hv}, P_{ov}, and P_{hd}) increase. At the same time, the increase in value of any parameter leads to an increase in overall channel utilization (see Fig. 1.4).

Another direction of research consists in an estimation of accuracy of the developed approximate formulae to calculate QoS metrics. Exact values (EV) of QoS metrics are determined from SGBE. It is important to note that under fulfillment of the mentioned assumptions related to the ratio of loading parameters of heterogeneous traffic the exact and approximate values (AV) are almost identical for all QoS metrics. Therefore, these comparisons are not shown here. At the same time, it is obvious that finding the EV of QoS metrics on the basis of the solution of SGBE appears effective only for models with a moderate dimension.

It is important to note the sufficiently high accuracy of the suggested formulae even for the case when the accepted assumption about the ratio of traffic loads is not fulfilled. To facilitate computational efforts, as EV of QoS metrics we use values that were calculated from explicit formulae, see [6], pages 131–135. In the mentioned work appropriate results are obtained for the special case where $b = 1$ and $\mu_v = \mu_d$. Let's note that condition $\mu_v = \mu_d$ contradicts our assumption $\mu_v >> \mu_d$. It is easy to execute a comparative analysis of results by means of Tables 1.1, 1.2, and 1.3 where the initial data are $N = 16$, $N_3 = 14$, $N_2 = 10$, $\lambda_{ov} = 10$, $\lambda_{hv} = 6$, $\lambda_{od} = 4$, $\lambda_{hd} = 3$, and $\mu_v = \mu_d = 2$.

Apparently from these tables, the highest accuracy of the developed approximate formulae is observed in calculation of QoS metrics for v-calls since for them the maximal difference between exact and approximate values does not exceed 0.001 (see Table 1.1). Small deviations take place in the calculation of QoS metrics for d-calls, but also here in the worst cases the absolute error of the proposed formulae does not exceed 0.09, which is perfectly reasonable for engineering practice (see Table 1.2). Similar results are observed for the average number of occupied channels of a cell (see Table 1.3). It is important to note that numerous numerical experiments have shown that at all admissible loads the accuracy of the proposed approximate formulae grows with growth in the total number of channels (i.e. with an increase in the dimension of state space of the model).

It is clear that in terms of simplicity and efficiency, the proposed approach is emphatically superior to the approach based on the use of balance equations for the calculation of QoS metrics of the given CAC in the model with non-identical channel occupancy time.

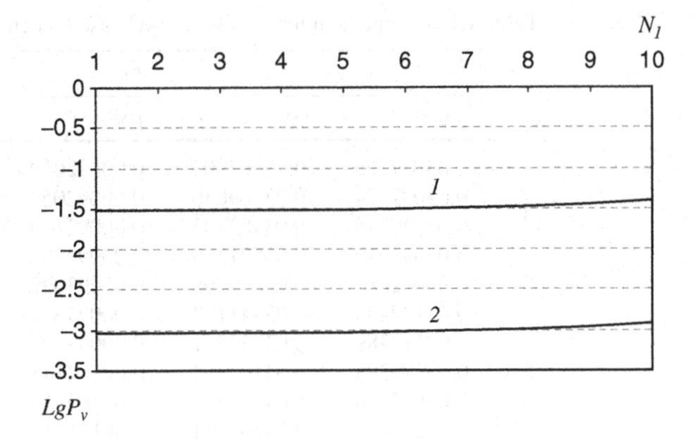

Fig. 1.2 Blocking probability of v-calls versus N_1: 1–P_{ov}; 2–P_{hv}

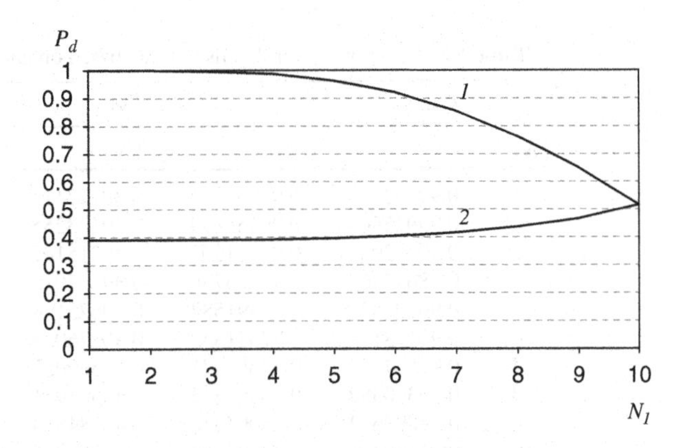

Fig. 1.3 Blocking probability of d-calls versus N_1: 1–P_{od}; 2–P_{hd}

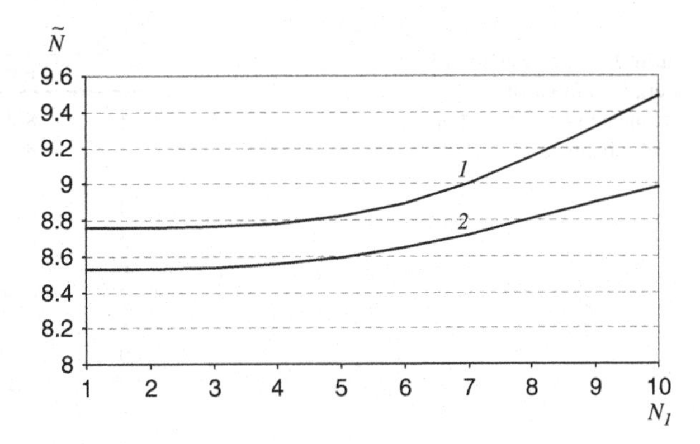

Fig. 1.4 Average number of busy channels versus N_1: 1–$N_3 = 15$; 2–$N_3 = 11$

Table 1.1 Comparison for v-calls in CAC based on guard channels

	P_{ov}		P_{hv}	
N_1	EV	AV	EV	AV
1	0.03037298	0.03465907	0.00092039	0.00119181
2	0.03037774	0.03469036	0.00092054	0.00119309
3	0.03040249	0.03482703	0.00092129	0.00119878
4	0.03048919	0.03521813	0.00092392	0.00121521
5	0.03072036	0.03604108	0.00093092	0.00125021
6	0.03122494	0.03741132	0.00094621	0.00130942
7	0.03217389	0.03932751	0.00097497	0.00139396
8	0.03377398	0.04168754	0.00102345	0.00150073
9	0.03627108	0.04432985	0.00109912	0.00162373
10	0.03997025	0.04706484	0.00121112	0.00175503

Table 1.2 Comparison for d-calls in CAC based on guard channels

	P_{od}		P_{hd}	
N_1	EV	AV	EV	AV
1	0.99992793	0.99985636	0.39177116	0.35866709
2	0.99925564	0.99855199	0.39183255	0.35886135
3	0.99612908	0.99271907	0.39215187	0.35969536
4	0.98645464	0.97565736	0.39327015	0.36203755
5	0.96398536	0.93891584	0.39625194	0.36685275
6	0.92198175	0.87621832	0.40276033	0.37462591
7	0.85564333	0.78660471	0.41500057	0.38506671
8	0.76370389	0.67487475	0.43563961	0.39731190
9	0.64880652	0.55004348	0.46784883	0.41028666
10	0.51556319	0.42295366	0.51556319	0.42295366

Table 1.3 Comparison for average number of busy channels in CAC based on guard channels

N_1	EV	AV
1	8.75786133	8.52991090
2	8.75908958	8.53136014
3	8.76473770	8.53753920
4	8.78196778	8.55476428
5	8.82125679	8.58985731
6	8.89293266	8.64583980
7	9.00241811	8.71992002
8	9.14705952	8.80533833
9	9.31596095	8.89429324
10	9.49204395	8.97976287

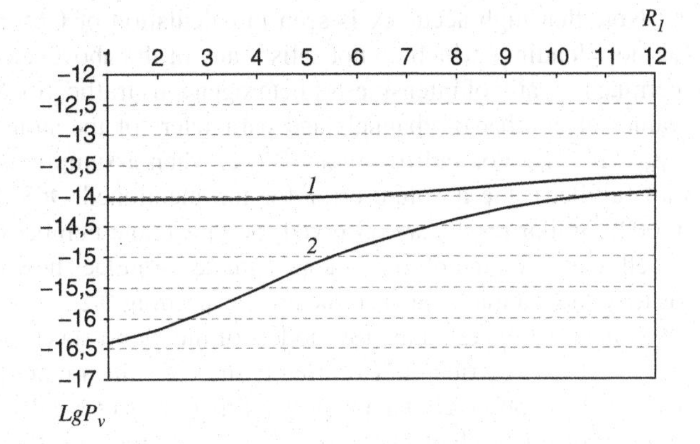

Fig. 1.5 Blocking probability of v-calls versus R_1; 1–P_{ov}, 2–P_{hv}

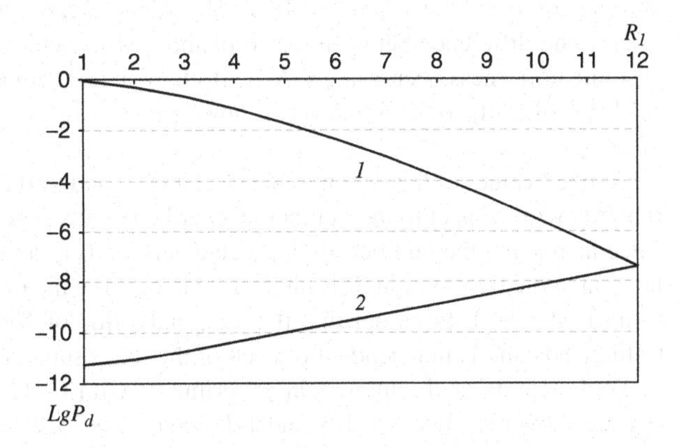

Fig. 1.6 Blocking probability of d-calls versus R_1; 1–P_{od}, 2–P_{hd}

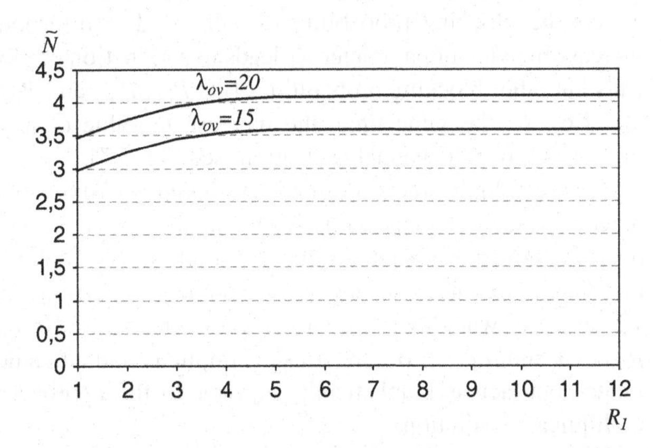

Fig. 1.7 Average number of busy channels versus R_1; 1–λ_{ov}=15, 2–λ_{ov}=20

Note that high accuracy is seen in calculation of QoS metrics for v-calls even at those loadings which do not satisfy any of the above-accepted assumptions concerning the ratio of intensities of heterogeneous traffic. So, for example, at the same values of number of channels and parameters of the strategy, at $\lambda_{ov}=4$, $\lambda_{hv}=3$, $\lambda_{od}=10$, $\lambda_{hd}=6$, and $\mu_v=\mu_d=2$ (i.e. when assumptions $\lambda_v >> \lambda_d$, $\mu_v >> \mu_d$ are not fulfilled), the absolute error for the mentioned QoS metric does not exceed 0.002. Similar results are observed for an average number of occupied channels of a cell. However, the proposed approximate formulae show low accuracy for d-calls as they show a maximal absolute error exceeding 0.2.

Numerical experiments for models of the mono-service CWN with guard channels were also carried out (for brevity these results are not presented here). As was expected, we observed that the probability of loss of h-calls (P_h) decreases when the number of guard channels increases. On the other hand, the probability of loss of o-calls (P_o) increases when the number of guard channels increases. And these probabilities are the same where $g=0$. These facts indicated that absolute fairness (in a sense the difference between loss probabilities) is achieved where $g=0$ whereas absolute unfairness occurs at $g=N–1$. Both P_h and P_o are increasing functions versus loads of traffic of different types, however, the two functions differ in their rates of change.

At fixed values of the total number of channels and traffic loads, $\tilde{N}$ is a decreasing function with respect to the number of guard channels. This fact was also expected, since increasing the number of guard channels leads to a worsening of channel utilization in the base station. It is interesting to observe the behavior of $\tilde{N}$ with respect to traffic loads. It turns out that the maximal value of $\tilde{N}$is defined by the sum of traffic loads and is independent of each of the loads separately.

Some results of the numerical experiments with CAC based on a cutoff strategy are shown in Figs. 1.5, 1.6, and 1.7, where $N=25$, $R_2=12$, $R_3=20$, $\lambda_{ov}=10$, $\lambda_{hv}=6$, $\lambda_{od}=4$, $\lambda_{hd}=3$, $\mu_v=10$, and $\mu_d=2$. As in CAC based on guard channels, the increase in value of one of the parameters (in an admissible area) favorably affects the blocking probability of calls of the corresponding type only. So, an increase in value of parameter R_1 leads to a reduction of blocking probability of od-calls but other blocking probabilities (i.e. P_{hv}, P_{ov}, and P_{hd}) increase (see Figs. 1.5 and 1.6). At the same time, the increase in value of any parameter leads to an increase in overall channel utilization (see Fig. 1.7).

The very high precision of the proposed approximate method should also be noted. Thus, in this case comparative analysis of the approximate results and the results obtained using a multiplicative solution (for small values of channels) shows that their differences are negligible. Some comparisons are shown in Tables 1.4, 1.5, and 1.6, where $N=16$, $R_3=14$, $R_2=12$, $\lambda_{ov}=10$, $\lambda_{hv}=6$, $\lambda_{od}=4$, $\lambda_{hd}=3$, $\mu_v=10$, and $\mu_d=2$. But in terms of simplicity and efficiency, the proposed approximate approach is emphatically superior to the approach based on the use of a multiplicative solution.

Numerical experiments with CAC based on individual pools of channels for heterogeneous calls were also carried out. For the sake of brevity these results are not shown here.

Table 1.4 Comparison for v-calls in CAC based on a cutoff strategy

	P_{ov}		P_{hv}	
R_1	EV	AV	EV	AV
1	1.63037E-09	5.06936E-09	4.08595E-10	3.83878E-09
2	1.89571E-09	7.46011E-09	1.28180E-09	6.55444E-09
3	7.04518E-09	1.36060E-08	6.64946E-09	1.28031E-08
4	3.58332E-08	2.56116E-08	3.55293E-08	2.48398E-08
5	1.66965E-07	4.49222E-08	1.66703E-07	4.41587E-08
6	6.58181E-07	7.03534E-08	6.57938E-07	6.95918E-08
7	2.15079E-06	9.75989E-08	2.15056E-06	9.68376E-08
8	5.79051E-06	1.21224E-07	5.79028E-06	1.20463E-07
9	1.28550E-05	1.37699E-07	1.28548E-05	1.36938E-07
10	2.36734E-05	1.46838E-07	2.36731E-05	1.46077E-07
11	3.65635E-05	1.50780E-07	3.65633E-05	1.50019E-07
12	4.64357E-05	1.51979E-07	4.64357E-05	1.51218E-07

Table 1.5 Comparison for d-calls in CAC based on a cutoff strategy

	P_{od}		P_{hd}	
R_1	EV	AV	EV	AV
1	8.38737E-01	8.37572E-01	3.77778E-08	3.84384E-09
2	4.35358E-01	4.62070E-01	4.50956E-08	6.56316E-09
3	2.45045E-01	1.98542E-01	7.25459E-08	1.28211E-08
4	1.17207E-01	6.78561E-02	1.53606E-08	2.48803E-08
5	1.68981E-02	1.87948E-02	4.04312E-08	4.42521E-08
6	8.97290E-03	4.32383E-03	1.17164E-08	6.98093E-08
7	4.27351E-03	8.46798E-04	3.30711E-08	9.73448E-08
8	1.82383E-04	1.44218E-04	8.44834E-07	1.21647E-07
9	7.00046E-05	2.17927E-06	1.90211E-06	1.39699E-07
10	2.43698E-05	3.04947E-06	3.80323E-06	1.52529E-07
11	7.83341E-06	4.96980E-07	7.00469E-06	1.65051E-07
12	8.05375E-07	1.86294E-07	9.15432E-06	1.86294E-07

Finally, we conducted a comparative analysis of the QoS metrics of two schemes: CAC based on the guard channels scheme and CAC based on the cutoff strategy. Comparison was performed over the broad range of number of channels and load parameters. In each access strategy the total number of channels is fixed and controllable parameters are N_1, N_2, N_3 (for CAC based on the guard channels scheme) and R_1, R_2, R_3 (for CAC based on the cutoff strategy). As is known, the behavior of QoS metrics with respect to the mentioned controllable parameters in different CAC is the same.

Some results of the comparison are shown in Figs. 1.8, 1.9, 1.10, 1.11, and 1.12 where labels 1 and 2 denote QoS metrics for CAC based on guard channels and CAC based on cutoff strategies, respectively. The input data are the same as for Tables 1.4, 1.5, and 1.6. In the graphs the parameter for CAC based on guard

Table 1.6 Comparison for average number of busy channels in CAC based on a cutoff strategy

	N_{av}	
R_1	EV	AV
1	2.783511	2.772190
2	3.146135	3.055078
3	3.310075	3.249941
4	3.351735	3.346218
5	3.395505	3.382321
6	3.400233	3.392265
7	3.409345	3.395523
8	3.411452	3.396039
9	3.411499	3.396129
10	3.411500	3.396142
11	3.411523	3.396144
12	3.411525	3.396145

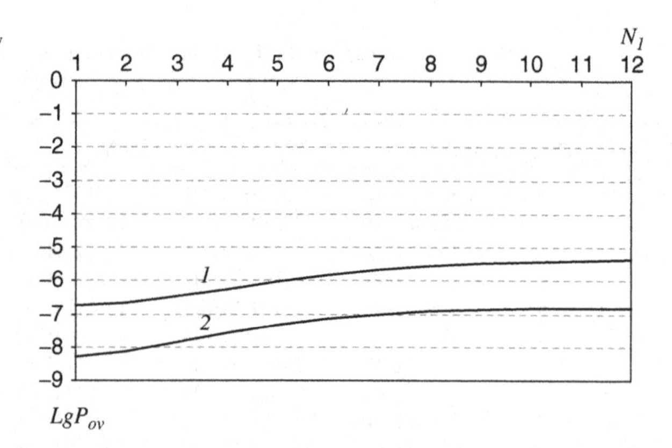

Fig. 1.8 Comparison for P_{ov} under different CAC

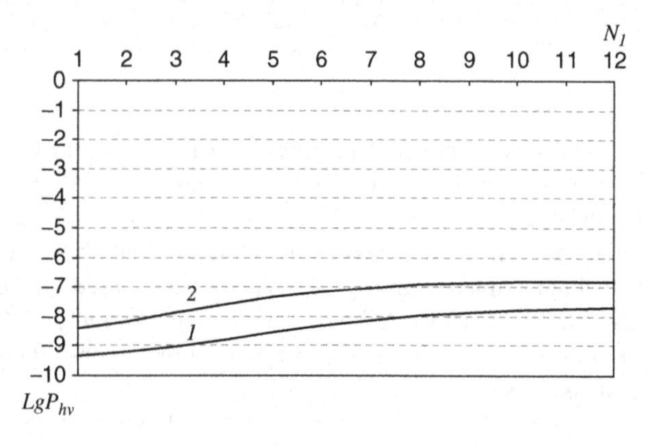

Fig. 1.9 Comparison for P_{hv} under different CAC

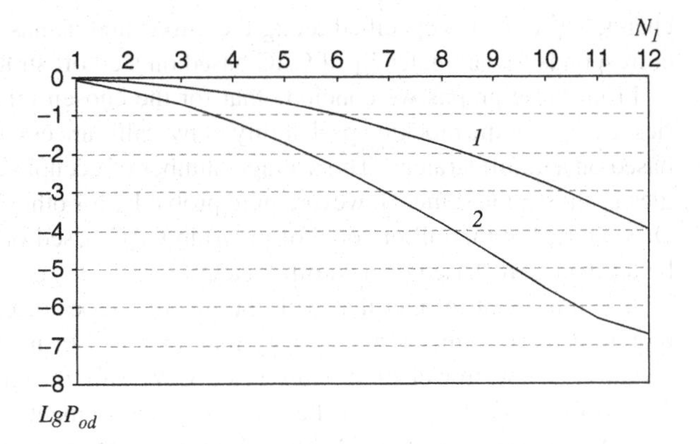

Fig. 1.10 Comparison for P_{od} under different CAC

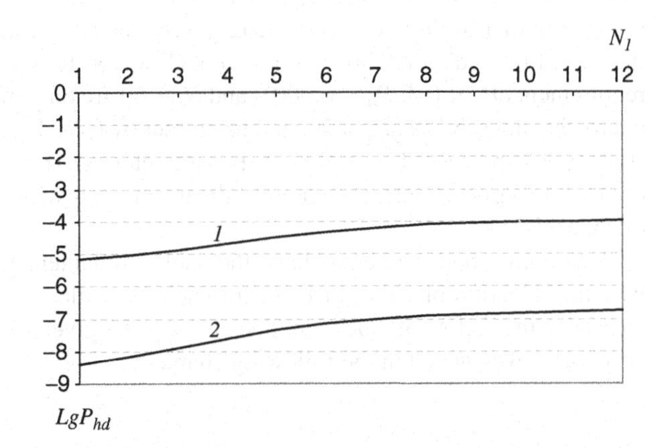

Fig. 1.11 Comparison for P_{hd} under different CAC

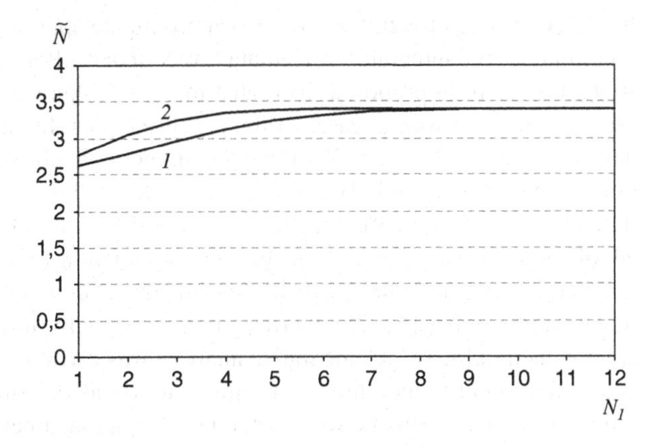

Fig. 1.12 Comparison for $\tilde{N}$ under different CAC

channels (i.e. N_1) is specified along the x-axis and as has been specified above, it corresponds to parameter R_1 of CAC based on a cutoff strategy.

From these graphs we conclude that for the chosen initial data three QoS metrics, except for the blocking probability of hv-calls, are essentially better under CAC based on a cutoff strategy. The average number of occupied channels in both strategies is almost the same. However, quite probably, for other values of initial data the QoS metrics (either all or some of them) in CAC based on guard channels will be better than CAC based on a cutoff strategy.

It is important to note that with the given number of channels, loads and QoS requirements of either CAC strategy may or may not meet the requirements. For instance, in the model of mono-service CWN for the given values of $N = 100$, ν_o=50 Erl, and $\nu_h = 35$ Erl the following requirements $P_o \leq 0.1$, $P_h \leq 0.007$, and $\tilde{N} \geq 80$ are not met with CAC based on guard channels irrespective of the value of parameter g (number of guard channels), whereas in CAC based on individual pools only that for h-calls meets the requirements at $r_h = 40$ (recall that r_h denotes the size of the individual pool for h-calls). However, for the same given initial data, requirements $P_o \leq 0.3$, $P_h \leq 0.0001$, and $\tilde{N} \geq 60$ are only met by CAC based on the guard channel scheme at $g = 20$, and never met by CAC based on an individual pool strategy irrespective of the value of its parameter r_h. Thus, it is possible to find an optimal (in a given context) strategy at the given loads without changing the number of channels.

Apparently, both strategies have the same implementation complexity. That is why the selection of either of them for each particular case must be based on the answer to the following question: does it meet the given QoS requirements? These issues are investigated in the following chapters.

1.4 Conclusion

In this chapter an effective and refined approximate approach to performance analysis of unbuffered integrated voice/data CWN under different multi-parametric CAC strategies has been proposed. Note that many well-known results related to mono-service CWN are special cases of the proposed ones. In almost all available works devoted to mono-service CWN the queuing model is investigated with the assumption that both handover and original calls are identical in terms of channel occupancy time (see [6, 7, 12, 14, 36, 38] and references therein). Since in known works models of mono-service CWN with different occupation times for o- and h-calls are not well investigated, the results of this chapter related to such networks are totally based on researches [22–25]. Corresponding results for integrated voice/data CWN are obtained in [8, 15]. Comparative analysis was carried out with the results of [6] and [12] in order to measure the accuracy of provided formulae. In [38], Chap. 11, a model wireless network with different occupation times for o- and h-calls was investigated. Here CAC with individual pools of channels for o- and h-calls and a common pool for both types of calls is investigated and a matrix-analytic method is used to provide a performance evaluation of the model with a finite queue of h-calls.

References

1. Akimaru H, Kawashima M (1993) Teletraffic theory and applications. Springer, London
2. Andreadis A, Giambene G (2003) Protocols for high efficiency wireless networks. Kluwer, Boston, MA
3. Boucherie RJ, Van Dijk NM (2000) On a queuing network model for cellular mobile telecommunications networks. Oper Res 48(1):38–49
4. Boucherie RJ, Mandjes M (1998) Estimation of performance measures for product form cellular mobile communications networks. Telecommun Syst 10:321–354
5. Casares-Giner V (2001) Integration of dispatch and interconnect traffic in a land mobile trunking system. Waiting time distributions. Telecommun Syst 10:539–554
6. Chen H, Huang L, Kumar S, Kuo CC (2004) Radio resource management for multimedia QoS supports in wireless networks. Kluwer, Boston, MA
7. Das Bit S, Mitra S (2003) Challenges of computing in mobile cellular environment – a survey. Comput Commun 26:2090–2105
8. Eom HY, Kim CS, Melikov AZ, Fattakhova MI (2009) Approximate method for QoS analysis of multi-threshold queuing model of multi-service wireless networks. In: Proceedings of 6th international workshop on simulation. St. Petersburg, Russia, pp 833–838
9. Freeman RL (1994) Reference manual for telecommunications engineering. Wiley, New York, NY
10. Gavish B, Sridhar S (1997) Threshold priority policy for channel assignment in cellular networks. IEEE Trans Comput 46(3):367–370
11. Greenberg AG, Srikant R, Whitt W (1999) Resource sharing for book-ahead and instantaneous-request calls. IEEE/ACM Trans Netw 7(1):10–22
12. Haring G, Marie R, Puigjaner R, Trivedi K (2001) Loss formulas and their application to optimization for cellular networks. IEEE Trans Veh Technol 50(3):664–673
13. Janevski T (2003) Traffic analysis and design of wireless IP networks. Artech House, Boston, MA
14. Katzela I, Naghshineh M (1996) Channel assignment schemes for cellular mobile telecommunication systems. IEEE Pers Commun, June: 10–31
15. Kim CS, Melikov AZ, Ponomarenko LA (2009) Numerical investigation of multithreshold access strategy in multiservice cellular wireless networks. Cybern Syst Anal 45(5):680–691
16. Kelly FP (1979) Reversibility and stochastic networks. Wiley, New York, NY
17. Korolyuk VS, Korolyuk VV (1999) Stochastic model of systems. Kluwer, Boston, MA
18. Korhonen J (2003) Introduction to 3G mobile systems. Artech House, Boston, MA
19. Li W, Chao X (2004) Modeling and performance evaluation of cellular mobile networks. IEEE/ACM Trans Netw 12(1):131–145
20. Lin YB, Chlamatac I (2000) Mobile networks. protocol and services. Wiley, New York, NY
21. Mehrotra A (1994) Cellular radio analogy and digital systems. Artech House, Norwood, MA
22. Melikov AZ, Babayev AT (2004) A new method of performance analysis of queuing model with guard channels. In: Proceedings of 5th international workshop on retrial queues, Seoul, Korea, pp 103–110
23. Melikov AZ, Babayev AT (2006) Refined approximations for performance analysis and optimization of queuing model with guard channels for handovers in cellular networks. Comput Commun 29(9):1386–1392
24. Melikov AZ, Fattakhova MI, Babayev AT (2004) Calculation and optimization of call processing procedures in cellular wireless communication networks. Autom Control Comput Sci 38(4):55–63
25. Melikov AZ, Fattakhova MI, Babayev AT (2005) Investigation of cellular communication networks with private channels for service of handover calls. Autom Control Comput Sci 39(3):61–69
26. Nanda S (1993) Teletraffic models for urban and suburban microcells: cell sizes and hand-off rates. IEEE Trans Veh Technol 42(4):673–682

27. Ogbonmwan SE, Wei L (2006) Multi-threshold bandwidth reservation scheme of an integrated voice/data wireless network. Comput Commun 29(9):1504–1515
28. Prasad R, Mohr W, Konhauser W (2000) 3G mobile communication systems. Artech House, Norwood, MA
29. Prasad R, Prasad S (2002) WLAN systems and wireless IP for next generation communications. Artech House, Norwood, MA
30. Prasad R, Ruggieri M (2003) Technology trends in wireless communications. Artech House, Boston, MA
31. Rappoport SS (1996) Wireless communications, principles and practice. Prentice-Hall, New York, NY
32. Redl SM, Weber MK, Oliphant MW (1995) An introduction to GSM. Artech House, Norwood, MA
33. Tekinay S, Jabbari B (1991) Handover policies and channel assignment strategies in mobile cellular networks. IEEE Commun Mag 29(11):42–46
34. Telephone Traffic Theory. Tables and Charts. Part 1. (1970) Siemens AG. Telephone and switching division, Munich
35. Schiller J (2000) Mobile communications. Addisson-Wesley, Norwood, MA
36. Sidi M, Starobinski D (1997) New call blocking versus handoff blocking in cellular networks. Wirel Netw 3:15–27
37. Wolff RW (1992) Poisson arrivals see time averages. Oper Res 30(2):223–231
38. Yue W, Matsumoto Y (2002) Performance analysis of multi-channel and multi-traffic on wireless communication networks. Kluwer, Boston, MA

Chapter 2
Performance Analysis of Call-Handling Processes in Buffered Cellular Wireless Networks

In this chapter effective numerical computational procedures to calculate QoS (Quality of Service) metrics of call-handling processes in mono-service Cellular Wireless Networks (CWN) with queues of either original (o-calls) or handover (h-calls) calls are proposed. Generalization of the results found here for integrated voice/data CWN is straightforward. Unlike classical models of mono-service CWN, here original and handover calls are assumed not to be identical in terms of time of radio channel occupancy. First we consider models of CWN with queues of h-calls in which for their prioritization a guard channels scheme is also used. We will then consider models of CWN with queues of o-calls and guard channels for h-calls. For both kinds of model the cases of limited and unlimited queues of patient and impatient calls are investigated. For the models with unlimited queues of heterogeneous calls the easily checkable ergodicity conditions are proposed. The high accuracy of the developed approximate formulae to calculate QoS metrics is shown.

2.1 Models with Queues for h-Calls

As mentioned in Chap. 1, the main method for prioritization of h-calls is the use of reserve channels (shared or isolated reservation). Another scheme for this purpose is the efficient arrangement of their queue in the base station. However, joint use of these schemes improves the QoS metrics of h-calls.

The required queue arrangement for h-calls can be realized in networks where microcells are covered by a certain macrocell, i.e. there exists a certain zone (handover zone – h-zone), within which mobile users can be handled in any of the neighboring cells. The time for the user to cross the h-zone is called the degradation interval. As the user enters the h-zone a check is made of the availability of free channels in a new cell. If a free channel exists, then the channel is immediately occupied and the h-procedure is considered to be successfully completed at the given stage; otherwise the given h-call continues to use the channel of the old (previous) cell while concurrently queuing for availability of a certain channel of a new cell. If the free channel does not appear in the new cell before completion of the degradation interval, then a forced call interruption of the h-call occurs.

L. Ponomarenko et al., *Performance Analysis and Optimization of Multi-Traffic on Communication Networks*, DOI 10.1007/978-3-642-15458-4_2,

Herein we consider models of four types: (i) limited queuing of h-call and infinite degradation interval; (ii) limited queuing of h-call and finite degradation interval; (iii) unlimited queuing of h-call and infinite degradation interval; (iv) unlimited queuing of h-call and finite degradation interval.

In all mentioned types of model it is assumed that a cell contains $N > 1$ radio channels and o-calls (h-calls) enter the given cell by the Poisson law with intensity λ_o (λ_h), the time of channel occupancy by o-calls (h-calls) being an exponentially distributed random quantity with a mean of μ_o^{-1} (μ_h^{-1}). Note, that the time of channel occupancy considers both components of occupancy time: the call service time and mobility. If during the service time of any type of call the h-procedure occurs, then due to the lack of memory of exponential distribution the remaining time of the given call service in a new cell (now the h-call) also has an exponential distribution with the same mean.

The different types of call are handled by the scheme of guard channels (shared reservation), i.e. an entering o-call is received only when there exists not less than $g+1$ free channels; otherwise the o-call is lost (blocked). A handover call is received with at least one free channel available; should all N channels be busy, the h-call joins the queue (limited or unlimited). In all model types it is assumed that at the moment a channel becomes free in a new cell the queue of h-calls (if it exists) is served by the FIFO (First-In-First-Out) process; with no queue the free channel stands idle.

2.1.1 Models with Finite Queues

First we consider the models of type (i), i.e. models with a limited queue of patient h-calls. In this model if all N channels are busy, then the entering h-call joins the queue with maximal size $B > 1$, if at least one vacant place is available; otherwise (i.e. when all places in the buffer are occupied) the h-call is lost. Since the degradation interval is infinite, the handover call cannot be lost should it be placed in a queue. In other words, the waiting h-calls are assumed to be patient. Note that this model is adequate for networks with slow velocity mobile users.

For a more detailed description of a cell's operation use is made of 2-D MC (Markov chain), i.e. the cell state at the arbitrary time instant is given by the vector $\mathbf{k} = (k_1, k_2)$, where k_i is the number of o-calls (h-calls) in the system, $i = 1,2$. Since o-calls are handled in a blocking mode and the system is conservative (i.e. with the queue available channel outages are not admitted), in the state $\mathbf{k}$ the number of h-calls in channels (k_2^s) and in the queue (k_2^q) are determined as follows:

$$k_2^s = \min\{N - k_1, k_2\}, k_2^q = (k_1 + k_2 - N)^+, \tag{2.1}$$

where $x^+ = \max(0, x)$. Therefore, the set of all possible states of the system is determined in the following way:

$$S := \{\mathbf{k}{:}k_1 = 0,1,\ldots,N-g;\ k_2 = 0,1,\ldots,N+B,\ k_1 + k_2^s \le N,\ k_2^q \le B\}. \tag{2.2}$$

Note 2.1. In the known works in view of the calls being identical in terms of channel occupation time, the state of a cell is described by a scalar magnitude which points out a general number of busy channels in a base station, i.e. as the mathematical model one applies 1-D MC. Since in the models studied the channel occupation time by different types of call is different, the description of a cell by a scalar magnitude is impossible in principle.

The elements of the generating matrix corresponding to 2-D MC, $q(\mathbf{k},\mathbf{k}')$, $\mathbf{k},\mathbf{k}' \in S$, are determined as follows (see Fig. 2.1):

$$q(\mathbf{k},\mathbf{k}') = \begin{cases} \lambda_o & \text{if } k_1 + k_2 \le N-g-1, \mathbf{k}' = \mathbf{k}+\mathbf{e}_1, \\ \lambda_h & \text{if } \mathbf{k}' = \mathbf{k}+\mathbf{e}_2, \\ k_1\mu_o & \text{if } \mathbf{k}' = \mathbf{k}-\mathbf{e}_1, \\ k_2^s\mu_h & \text{if } \mathbf{k}' = \mathbf{k}-\mathbf{e}_2, \\ 0 & \text{in other cases.} \end{cases} \tag{2.3}$$

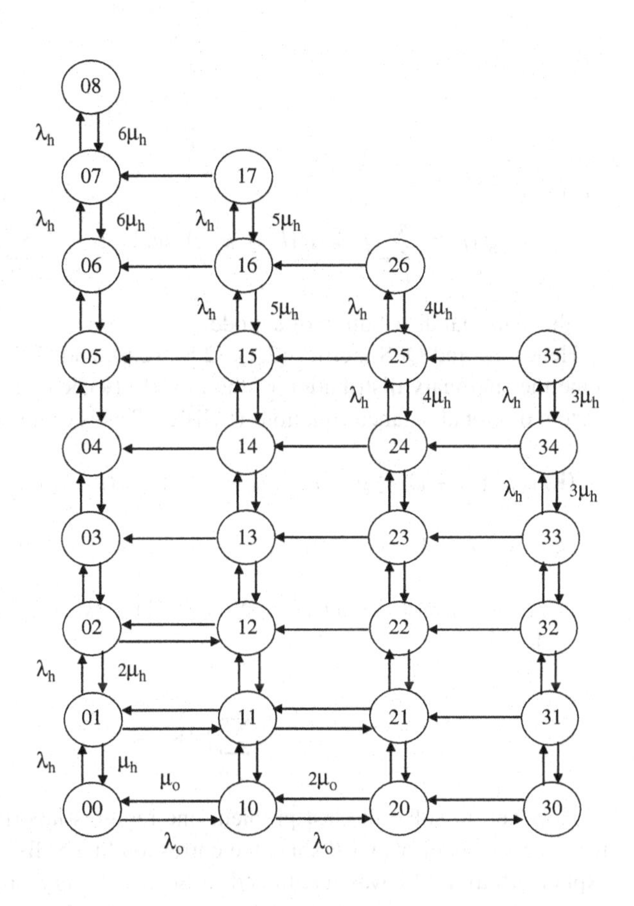

Fig. 2.1 State transition diagram for the model with a limited queue of patient h-calls, $N = 6$, $g = 3$, $B = 2$

Hence, the mathematical model of the given system is presented by 2-D MC with a state space (2.2) for which the elements of the generating matrix are determined by means of relations (2.3).

The stationary probability of state $\mathbf{k}$ is denoted by $p(\mathbf{k})$. Then in view of the model being Markovian, according to the PASTA theorem we find that the dropping probability of h-calls (P_{h}) and probability of blocking of o-calls (P_{o}) are determined in the following way:

$$P_{\mathrm{h}} := \sum_{k \in S} p(\mathbf{k}) \delta\left(k_2^{\mathrm{q}}, B\right), \tag{2.4}$$

$$P_{\mathrm{o}} := \sum_{k \in S} p(\mathbf{k}) I(k_1 + k_2^{\mathrm{s}} \geq N - g). \tag{2.5}$$

The average number of busy channels of the cell ($\tilde{N}$) and the average length of the queue of h-calls (L_{h}) are also determined via stationary distribution of the model:

$$\tilde{N} := \sum_{j=1}^{N} j \varsigma\,(j), \tag{2.6}$$

$$L_{\mathrm{h}} := \sum_{l=1}^{B} l \tau\,(l), \tag{2.7}$$

where

$$\varsigma\,(j) := \sum_{k \in S} p\,(\mathbf{k})\, \delta\left(k_1 + k_2^{\mathrm{s}}, j\right) \text{ and } \tau\,(l) := \sum_{k \in S} p\,(\mathbf{k})\, \delta\left(k_2^{\mathrm{q}}, l\right)$$

are the marginal distribution of a model.

Hence, to find QoS metrics (2.4), (2.5), (2.6), and (2.7) it is necessary to determine the stationary distribution of the model $p(\mathbf{k}), \mathbf{k} \in S$, from the corresponding system of global balance equations (SGBE). This is of the following form:

$$\begin{aligned} & p\,(\mathbf{k}) \left(\lambda_{\mathrm{o}} I\left(k_1 + k_2^{\mathrm{s}} \leq g - 1\right) + \lambda_{\mathrm{h}} \left(1 - \delta\left(k_2^{\mathrm{q}}, B\right) + \left(k_1 + k_2^{\mathrm{s}}\right) \mu\right)\right) \\ & = \lambda_{\mathrm{o}} p\,(\mathbf{k} - \mathbf{e}_1) \left(1 - \delta\,(k_1, 0)\right) I\left(k_1 + k_2^{\mathrm{s}} \leq g - 2\right) + \lambda_{\mathrm{h}} p\,(\mathbf{k} - \mathbf{e}_2) \left(1 - \delta\,(k_2, 0)\right) \\ & \quad + (k_1 + 1)\, \mu p\,(\mathbf{k} + \mathbf{e}_1)\, I\,(k_1 < N - g - 1) + (k_2^{\mathrm{s}} + 1) \mu p\,(\mathbf{k} + \mathbf{e}_2)\, I\left(k_2^{\mathrm{s}} < N\right), \\ & \quad \mathbf{k} \in S, \end{aligned} \tag{2.8}$$

$$\sum_{k \in S} p\,(\mathbf{k}) = 1 \tag{2.9}$$

However, to solve the last problem one requires laborious computation efforts for large values of N and B since the corresponding SGBE (2.8) and (2.9) have no explicit solution. As was mentioned in Sect. 1.2, very often the solution of such problems is evident if the corresponding MC has a reversibility property and hence there exists a stationary distribution for it of the multiplicative type. However, by applying Kolmogorov criterion for 2-D MC one can easily demonstrate that the given MC is not reversible. Indeed, the necessary reversibility condition states that if the transition from state (i, j) into the state (i', j') exists, then the reverse transition also exists. However, for the MC considered this condition is not fulfilled. So by the relations (2.3) in the given MC the transition $(k_1, k_2) \to (k_1 - 1, k_2)$ with intensity $k_1\mu_o$ exists, where $k_1 + k_2 > N-g$, while the inverse transition does not exist.

To overcome these difficulties one suggests employing the approximate method of calculating the stationary distribution of the 2-D MC. It is acceptable for models of micro- and picocells for which the intensity of h-calls entering greatly exceeds that of o-calls and the talk time generated by an h-call is short. In other words, here it is assumed that $\lambda_h >> \lambda_o$, $\mu_h >> \mu_o$ (for respective comments see Sect. 1.2).

Consider the following splitting of the state space (2.2):

$$S = \bigcup_{j=0}^{N-g} S_j, \quad S_j \bigcap S_{j'} = \varnothing, \; j \neq j', \tag{2.10}$$

where

$$S_j := \{\mathbf{k} \in S : k_1 = j\}, j = \overline{0, N-g}.$$

The sets S_j are combined in merged state $<j>$ and the merging function with the domain (2.2) is introduced:

$$U(\mathbf{k}) = <j> \;\; \text{if}\; \mathbf{k} \in S_j, \; j = \overline{0, N-g}. \tag{2.11}$$

The merging function (2.11) specifies the merged model, which is 1-D MC with state space $\tilde{S} := \left\{<j>: j = \overline{0, N-g}\right\}$.

The above assumption about relations of loading parameters of different call types provides the fulfillment of conditions necessary for correct application of phase-merging algorithms: intensities of transitions between states inside each class S_j, $j = 0, 1, \ldots, N-g$, essentially exceed intensities of transitions between states from different classes.

To find the stationary distribution of the initial model one needs a preliminary determination of the stationary distribution of split and merged models. The stationary distribution of the jth split model with space of states S_j is denoted by $\rho^j(i), j = 0, 1, \ldots, N-g, \; i = 0, 1, \ldots, N+B-j$, i.e. $\rho^j(i)$ is the stationary probability of state $(j, i) \in S_j$ in jth split model. It is determined as the stationary distribution of classical queuing system *M/M/N–j/B* with the load ν_h Erl, i.e.

$$\rho^j(i) = \begin{cases} \dfrac{\nu_h^i}{i!}\rho^j(0) \text{ if } i = \overline{1, N-j}, \\ \dfrac{\nu_h^i}{(N-j)!(N-j)^{i+j-N}}\rho^j(0) \text{ if } i = \overline{N-j+1, N-j+B}, \end{cases} \tag{2.12}$$

where

$$\rho^j(0) = \left(\sum_{i=0}^{N-j} \frac{v_h^i}{i!} + \frac{1}{(N-j)!} \sum_{i=N-j+1}^{N-j+B} \frac{v_h^i}{(N-j)^{i+j-N}} \right)^{-1}.$$

To find the stationary distribution of the merged model one should preliminarily determine the elements of the generating matrix corresponding to 1-D MC denoted by $q(<j', <j''>), <j'>, <j''> \in \tilde{S}$. The following relations determine the mentioned parameters:

$$q(<j'>, <j''>) = \begin{cases} \lambda_o \cdot \Lambda(j'+1) & \text{if } j'' = j'+1, \\ j'\mu & \text{if } j'' = j'-1, \\ 0 & \text{in other cases,} \end{cases} \tag{2.13}$$

where

$$\Lambda(i+1) = \rho^i(0) \sum_{j=0}^{N-g-i-1} \frac{v_h^j}{j!},\ i = \overline{0, N-g-1}.$$

The last relations imply the stationary distribution of the merged model $\pi(<j>), <j> \in \tilde{S}$, to be determined as the corresponding distribution of the appropriate birth-and-death process. In other words

$$\pi(<j>) = \frac{v_o^j}{j!} \prod_{i=1}^{j} \Lambda(i) \pi(<0>),\ j = \overline{1, N-g}, \tag{2.14}$$

where

$$\pi(<0>) = \left(1 + \sum_{i=1}^{N-g} \frac{v_o^i}{i!} \prod_{j=1}^{i} \Lambda(j) \right)^{-1}.$$

Summing up everything stated above one can offer the following approximate formulae for calculating the QoS metrics (2.4), (2.5), (2.6), and (2.7) of the given model:

$$P_h \approx \sum_{j=0}^{N-g} \rho^j(N+B-j) \pi(<j>); \tag{2.15}$$

$$P_o \approx \sum_{j=0}^{N-g} \sum_{i=N-g-j}^{N+B-j} \rho^j(i) \pi(<j>); \tag{2.16}$$

$$\tilde{N} \approx \sum_{j=1}^{N-g} j \sum_{i=0}^{j} \rho^i (j-i)\pi(<i>)$$
$$+ \sum_{j=N-g+1}^{N-1} j \sum_{i=0}^{N-g} \rho^i (k-i)\pi(<i>) + N \sum_{i=0}^{N-g} \sum_{j=N-i}^{N+B-i} \rho^i (j)\pi(<i>); \tag{2.17}$$

$$L_h \approx \sum_{j=1}^{N+B-g} j \sum_{i=0}^{g} \rho^i (j)\pi(<i>) + \sum_{j=N+B-g+1}^{N+B} j \sum_{i=0}^{N+B-j} \rho^i (j)\pi(<i>). \tag{2.18}$$

Now we can generalize the obtained results to the model with a limited queue of h-calls with a finite degradation interval [i.e. to the model of type (ii)]. Unlike the previous model in this model it is assumed that the degradation interval for h-calls are independent, equally distributed random quantities having exponential distribution with the finite mean γ^{-1}.

The state space of the given model is defined by (2.2) also. But for the given model the elements of the generating matrix corresponding to 2-D MC, $q(\mathbf{k}, \mathbf{k}')$, $\mathbf{k}, \mathbf{k}' \in S$ are determined as follows (see Fig. 2.2):

$$q(\mathbf{k}, \mathbf{k}') = \begin{cases} \lambda_o & \text{if } k_1 + k_2 < N - g - 1,\ \mathbf{k}' = \mathbf{k} + \mathbf{e_1}, \\ \lambda_h & \text{if } \mathbf{k}' = \mathbf{k} + \mathbf{e_2}, \\ k_1 \mu & \text{if } \mathbf{k}' = \mathbf{k} - \mathbf{e_1}, \\ k_2^s \mu \delta(k_2{}^q, 0) + (k_2{}^s \mu + k_2{}^q \gamma)(1 - \delta(k_2{}^q, 0)) & \text{if } \mathbf{k}' = \mathbf{k} - \mathbf{e_2}, \\ 0 & \text{in other cases.} \end{cases} \tag{2.19}$$

The stationary probability of blocking of o-calls (P_o) in this model is determined also by the formula (2.5). However, in this model losses of h-calls occur in the following cases:

(a) If at the moment of its entering the queue there are already B calls of the given type;
(b) If the interval of its degradation is completed before it gains access to a free channel.

Hence the given QoS metric is determined as follows:

$$P_h := \sum_{k \in S} p(\mathbf{k}) \delta\left(k_2^q, B\right) + \frac{1}{\lambda_h} \sum_{i=1}^{B} i\gamma \sum_{j=0}^{N-g} p(j, N+i-j). \tag{2.20}$$

In the last formulae the first term of the sum defines the probability of the event corresponding to case (a) above, while the second term of the sum defines the probability of even corresponding to case (b) above.

For the given model the average number of busy channels of the cell ($\tilde{N}$) and the average length of the queue of h-calls (L_h) are also determined analogously to

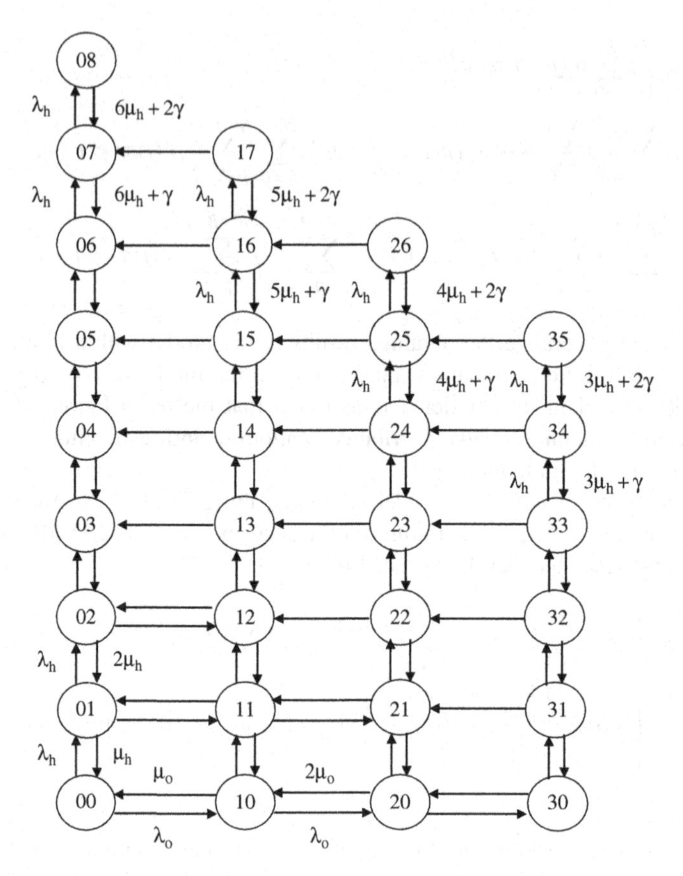

Fig. 2.2 State transition diagram for the model with a limited queue of impatient h-calls, $N = 6$, $g = 3$, $B=2$

(2.6) and (2.7). And SGBE for the given model is also derived in a similar way to (2.8) and (2.9). However, as in the model with patient h-calls, to solve the indicated SGBE requires laborious computation efforts for large values of N and B since the corresponding SGBE has no explicit solution. But to overcome these difficulties use is made of the approximate method proposed above for the model with patient h-calls.

In this case the splitting (2.10) of state space (2.2) is considered also and by function (2.11) an appropriate merged model with state space $\tilde{S} := \{< j >: j = \overline{0, N-g}\}$ is constructed. But in this case the stationary distribution in the jth split model is determined as follows:

$$\rho^j(i) = \begin{cases} \dfrac{v_h^i}{i!}\rho^j(0) & \text{if } i = \overline{1, N-j}, \\ \dfrac{v_h^{N-j}}{(N-j)!}\displaystyle\prod_{l=1}^{i+j-N}\dfrac{\lambda_h^l}{(N-j)\,\mu + l\gamma}\rho^j(0) & \text{if } i = \overline{N-j+1, N-j+B}, \end{cases} \tag{2.21}$$

where

$$\rho^j(0) = \left(\sum_{i=0}^{N-j} \frac{v_h^i}{i!} + \frac{v_h^{N-j}}{(N-j)!} \sum_{i=N-j+1}^{N-j+B} \prod_{l=1}^{i+j-N} \frac{\lambda_h^l}{(N-j)\,\mu + l\gamma} \right)^{-1}, j = \overline{0, N-g}.$$

Furthermore, the stationary distribution of the merged model $\pi(<j>)$, $<j>\in \tilde{S}$ is determined in a similar way to (2.14). Note that in this case the terms in (2.13) and (2.14) are calculated by taking into account formulae (2.21).

Making use of the above results and omitting the known intermediate mathematical calculations the following formulae to calculate the QoS metrics of a network with a limited queue and finite interval of degradation of h-calls are obtained:

$$P_h \approx \sum_{i=0}^{N-g} \rho^i(N+B-i)\pi(i) + \frac{1}{\lambda_h} \sum_{i=1}^{B} i\gamma \sum_{j=0}^{N-g} \rho^j\,(N+i-j)\,\pi\,(j); \tag{2.22}$$

$$P_o \approx \sum_{j=0}^{N-g} \sum_{i=N-g-j}^{N+B-j} \rho^j(i)\pi(<j>); \tag{2.23}$$

$$\begin{aligned} \tilde{N} &\approx \sum_{j=1}^{N-g} j \sum_{i=0}^{j} \rho^i(j-i)\pi(<i>) \\ &+ \sum_{j=N-g+1}^{N-1} j \sum_{i=0}^{N-g} \rho^i(j-i)\pi(<i>) + N \sum_{j=0}^{N-g} \pi(<j>) \sum_{i=N-j}^{N+B-j} \rho^j(i); \end{aligned} \tag{2.24}$$

$$L_h \approx \sum_{j=1}^{N+B-g} j \sum_{i=0}^{g} \rho^i(j)\pi(<i>) + \sum_{j=N+B-g+1}^{N+B} j \sum_{i=0}^{N+B-j} \rho^i(j)\pi(<i>). \tag{2.25}$$

2.1.2 Models with Infinite Queues

We now consider the model of type (iii), i.e. a model of a cell with an unlimited queue of h-calls and an infinite degradation interval. The set of all possible states of the given model is determined in the following way:

$$S := \left\{ \mathbf{k}{:}k_1 = 0, 1, \ldots, N-g;\; k_2 = 0, 1, \ldots;\; k_1 + k_2^s \le N \right\}. \tag{2.26}$$

The number of h-calls in the queue and in channels and the elements of the generating matrix are calculated in a similar way to (2.1) and (2.3), respectively (see Fig. 2.3).

The required QoS metrics in this case is also calculated via a stationary distribution of the model that is determined from the corresponding SGBE of infinite

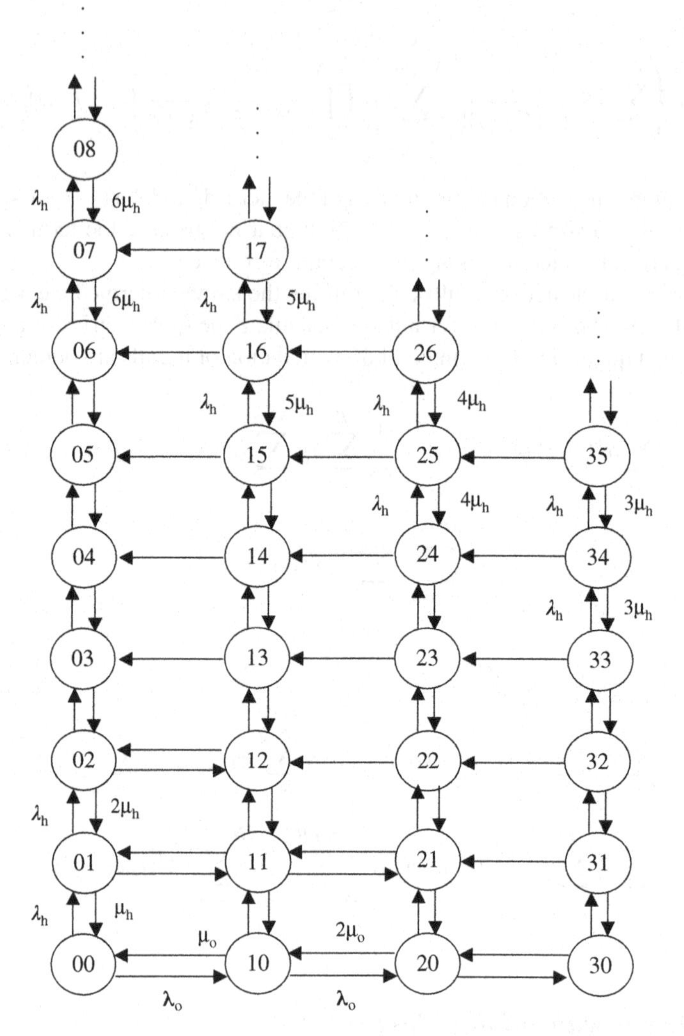

Fig. 2.3 State transition diagram for the model with an unlimited queue of patient h-calls, $N=6$, $g=3$

dimension. However, the employment of the method of two-dimensional generating functions for finding the stationary distribution of the given model from the mentioned SGBE is related to well-known computational and methodological difficulties. In relation to this we shall apply the above-described approach to calculating the stationary distribution of the model.

Without repeating the above procedures we just note that here we also made use of the split scheme of the state space (2.26) analogous to (2.10). Since the selection of the split scheme completely specifies the structures of the split and merged models, below only minor comments are made on the formulae suggested.

The stationary distribution of the *j*th split model is determined as the stationary distribution of the classical queuing system *M/M/N–j/*∞ with the load ν_h Erl, i.e.

$$\rho^j(i) = \begin{cases} \dfrac{\nu_h{}^i}{i!}\rho^j(0), & i = \overline{1, N-j}, \\ \dfrac{\nu_h{}^i (N-j)^{N-j-i}}{(N-j)!}\rho^j(0), & i \geq N-j, \end{cases} \tag{2.27}$$

where

$$\rho^j(0) = \left(\sum_{i=0}^{N-j-1} \frac{\nu_h{}^i}{i!} + \frac{\nu_h{}^{N-j}}{(N-j)!} \cdot \frac{N-j}{N-j-\nu_h} \right)^{-1}, \; j = 0, 1, \ldots, N-g.$$

The ergodicity condition of the *j*th split model is $\nu_h < N–j$. Hence, for the stationary mode to exist in each split model the condition $\nu_h < g$ should be fulfilled. Note that the model ergodicity condition is independent of the o-calls load. This should have been expected since by the assumption $\lambda_h >> \lambda_o$, $\mu_h >> \mu_o$ o-calls are handled by the scheme with pure losses. In the particular case $g = 1$ we obtain the condition $\nu_h < 1$.

The stationary distribution of the merged model is $\pi(<j>), <j> \in \tilde{S}$ determined in a similar way to (2.14). However, in this case one should consider the fact that in the above formulae the parameters $\rho^j(0)$, $j = 0, 1, \ldots, N-g-1$, are calculated from (2.27).

After performing the necessary mathematical transformations one obtains the following approximate formulae for calculating the QoS metrics of the model with an unlimited queue of patient h-calls and guard channels available:

$$P_o \approx 1 - \sum_{j=0}^{N-g-1} \sum_{i=0}^{N-g-1-j} \rho^j(i)\,\pi(<j>); \tag{2.28}$$

$$\tilde{N} \approx \sum_{j=1}^{N-g} j \sum_{i=0}^{j} \rho^i(j-i)\pi(<i>) + \sum_{j=N-g+1}^{N-1} j \sum_{i=0}^{N-g} \rho^i(k-i)\pi(<i>)$$
$$+ N \sum_{j=0}^{N-g} \pi(<j>) \left(1 - \sum_{i=0}^{N-j-1} \rho^j(i)\right); \tag{2.29}$$

$$L_h \approx \sum_{i=0}^{N-g} \pi(<i>)\,\rho^i(0)\,\frac{\nu_h^{N+1-i}}{(N-i)!} \cdot \frac{N-i}{(N-i-\nu_h)^2}. \tag{2.30}$$

Now we consider the model of type (iv), i.e. a model with an unlimited queue of impatient h-calls with the guard channels available. In the given model the impatient h-call can be lost from the unlimited queue unless a single channel of a new cell is

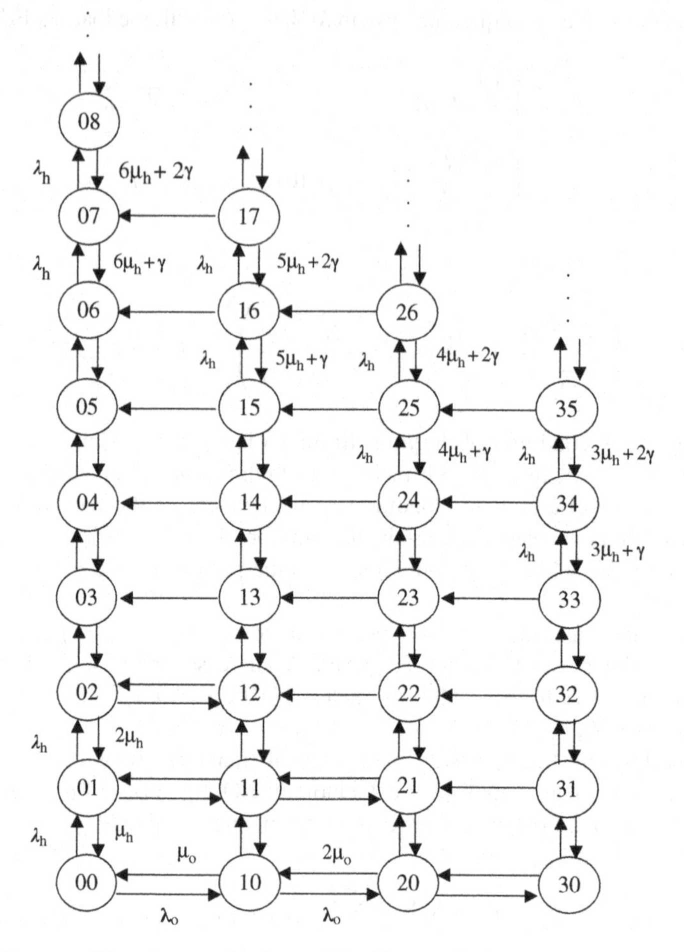

Fig. 2.4 State transition diagram for the model with an unlimited queue of impatient h-calls, $N = 6$, $g = 3$

freed before termination of its degradation interval. To obtain tractable results as in the model of type (ii) it is assumed that degradation intervals for all h-calls are independent, equally exponentially distributed random quantities with the finite mean γ^{-1}. The state space of this model is given by means of the set (2.26). However, thereby the elements of the generating matrix of the corresponding 2-D MC are determined in a similar way to (2.19) (see Fig. 2.4). Similarly to the previous model for the given model one can develop the appropriate SGBE for stationary probabilities of the system states. However, the above-mentioned difficulties of applying such SGBE in this case are even more complicated since to solve this one the approximate approach is used.

The stationary distribution of the jth split model in the given case is determined by

$$\rho^j(i) = \begin{cases} \dfrac{\nu_h^{\,i}}{i!}\rho^j(0), & \text{if } i = \overline{1, N-j}, \\ \dfrac{\nu_h^{\,N-j}}{(N-j)!}\displaystyle\prod_{k=N-j+1}^{i} \dfrac{\lambda_h}{(N-j)\,\mu_h + (k+j-N)\,\gamma}\rho^j(0), & \text{if } i \geq N-j+1, \end{cases} \tag{2.31}$$

where

$$\rho^j(0) = \left(\sum_{i=0}^{N-j} \frac{\nu_h^{\,i}}{i!} + \frac{\nu_h^{\,N-j}}{(N-j)!} \sum_{m=N-j+1}^{\infty} \prod_{k=N-j+1}^{m} \frac{\lambda_h}{(N-j)\,\mu_h + (k+j-N)\,\gamma} \right)^{-1}.$$

It is worth noting that in the given model at any permissible values of load parameters in the system there exists a stationary mode. It can be easily proved since the analysis of the ratio limit of two neighboring terms of a series shows that the numerical series

$$R := \sum_{m=N-j+1}^{\infty} \prod_{k=N-j+1}^{m} \frac{\lambda_h}{(N-j)\,\mu_h + (k+j-N)\,\gamma} \tag{2.32}$$

involved in determining $\rho^j(0)$ [see formulae (2.31)] converges at any positive values of load parameters of h-calls and degradation interval. However, unfortunately one does not manage to find the exact value of the sum of series (2.32) but we manage to find the following limits of this sum change:

$$\exp\left(\frac{\lambda_h}{(N-j)\,\mu_h + \gamma}\right) - 1 \leq R \leq \exp\left(\frac{\lambda_h}{\gamma}\right) - 1. \tag{2.33}$$

Note 2.2. From the last relations one can see that in a particular case, when $(N-j)\,\mu_h << \gamma$ or the quantity $(N-j)\,\mu_h$ is sufficiently small, the approximate value of sum R can be used as the right-hand side of inequality (2.33).

The QoS metrics of the given model are determined like (2.28), (2.29), and (2.30) but now the described formulae involve the corresponding distributions for the model with impatient h-calls. In the given model losses of h-calls from the queue due to their impatience also occur. The probability of such an event is calculated as follows:

$$P_h = \frac{1}{\lambda_h} \sum_{n=1}^{\infty} n\gamma \sum_{i=0}^{N-g} p(i, N+n-i)\,. \tag{2.34}$$

2.1.3 Numerical Results

The algorithms suggested allow one to study the behavior of QoS metrics of the systems investigated in all admissible ranges of changes to their structural and load parameters. In order to be short, among the models of cells with a queue of h-calls, only the results for models with unlimited queues are given in detail.

Some results of numerical experiments for the model with patient h-calls at $N = 15$ and $\mu_o = 0.5$ are shown in Figs 2.5, 2.6, and 2.7. They completely confirmed all theoretical expectations. So, the probability of losing o-calls grows (see Fig. 2.5) and the average number of busy channels (see Fig. 2.6) and the average number of h-calls in the queue (see Fig. 2.7) falls as the number of guard channels increases. As changes to the average time of h-call delay coincide with the analogous dependence on their average number in a queue, the details of this function are not presented here.

Note that all the functions under study are increasing with respect to intensity of o-call traffic. However, unlike the function $\tilde{N}$ the rates of change of the functions P_o and L_h are sufficiently high. So, at $N = 15$, $\nu_o = 4$, and $\nu_h = 0.8$ the values of functions P_o and L_h at the points $g = 1$ and $g = 10$ equal $P_o(1) = 3.8\text{E-}04$, $P_o(10) = 2.8\text{E-}01$, $L_h(1) = 9.9\text{E-}05$, $L_h(10) = 1.2\text{E-}10$, and those of the function $\tilde{N}$ at these points equal 4.7985 and 3.6551.

The analysis of results of numerical experiments shows that regardless of the essential difference in loads of o-calls, as the number of guard channels grows the corresponding values of functions P_o and L_h become closer. For example, in two experiments at $N = 15$ the load parameters were selected in this way: (1) $\nu_o = 4$, $\nu_h = 0.8$; (2) $\nu_o = 2$, $\nu_h = 0.75$. For this data the following ratios hold: $P_o^1(1)/P_o^2(1) \approx 400$, $P_o^1(14)/P_o^2(14) \approx 1.01$; $L_h^1(1)/L_h^2(1) \approx 10^3$, $L_h^1(14)/L_h^2(14) \approx 3$,

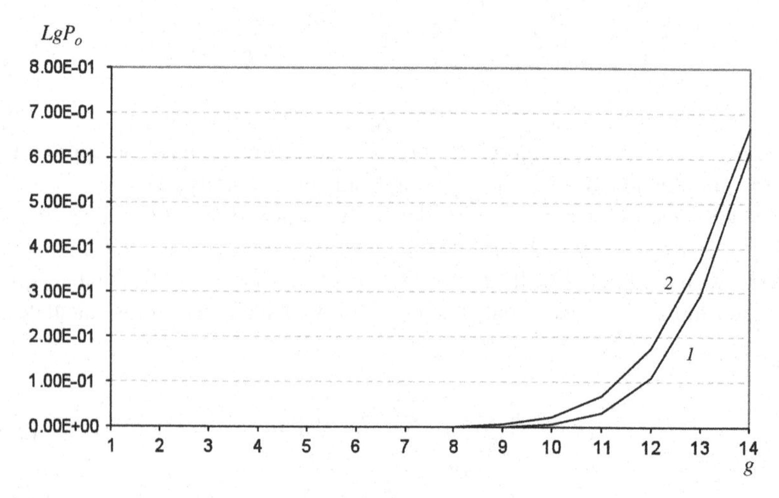

Fig. 2.5 P_o versus g for the model with an unlimited queue of h-calls in the case where $N = 15$; $\mu_o = 0.5$; $1 - \lambda_o = 2$, $\lambda_h = 4$ $\mu_h = 5$; $2 - \lambda_o = 1$, $\lambda_h = 3$ $\mu_h = 4$

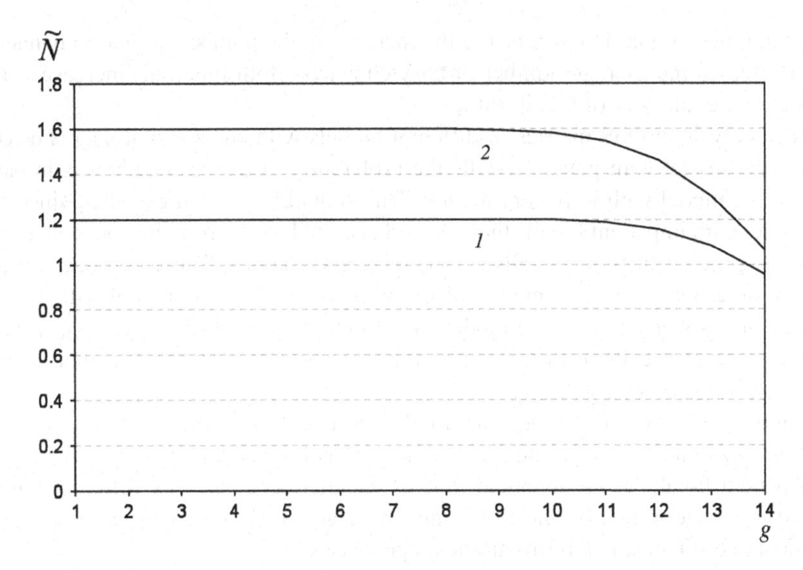

Fig. 2.6 $\tilde{N}$ versus g for the model with an unlimited queue of h-calls in the case where $N = 15$; $\mu_o = 0.5$; $1 - \lambda_o = 2$, $\lambda_h = 4$ $\mu_h = 5$; $2 - \lambda_o = 1$, $\lambda_h = 3$ $\mu_h = 4$

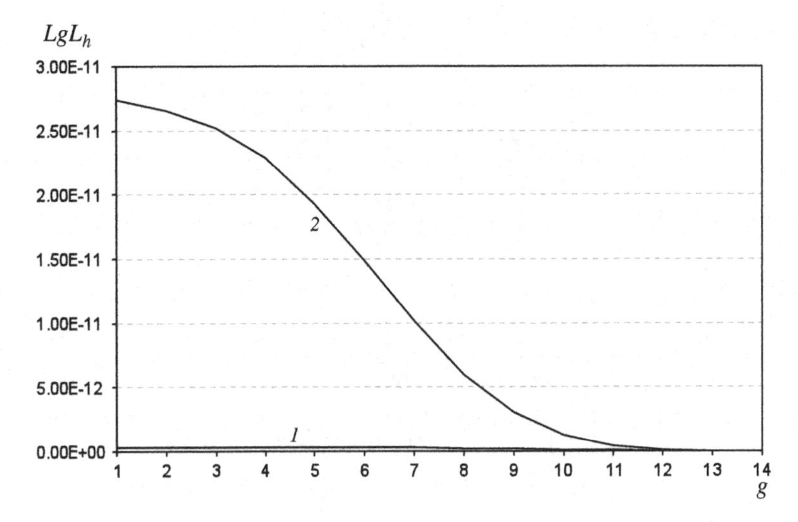

Fig. 2.7 L_h versus g for the model with an unlimited queue of h-calls in the case where $N = 15$; $\mu_o = 0.5$; $1 - \lambda_o = 2$, $\lambda_h = 4$ $\mu_h = 5$; $2 - \lambda_o = 1$, $\lambda_h = 3$ $\mu_h = 4$

where P_o^i (L_h^i) is the value of the function P_o (L_h) in the ith experiment, $i = 1,2$. The corresponding ratios for the function $\tilde{N}$are of the form $\tilde{N}^1(1)/\tilde{N}^2(1) \approx 1.8, \tilde{N}^1(14)/\tilde{N}^2(14) \approx 1.1$.

The results of numerical experiments for the model with impatient h-calls showed that the probability of losing o-calls also grew as the number of guard channels increased. In this model the probability of losing h-calls decreases with respect to the number of guard channels and the increase in the number of guard channels also decreases the average number of busy channels. Both functions increase with respect to the intensity of o-call traffic.

The analysis of QoS metrics of different models with equal initial data showed that in the model with patient h-calls the probability of losing o-calls was higher than in the model with impatient h-calls. This should have been expected since in the model with impatient h-calls there occur losses of h-calls from the queue thereby increasing the chances for o-calls to occupy a free channel. These comments also refer to other QoS metrics namely utilization of channels in the model with impatient h-calls is worse than in the model with patient h-calls. And the average length of the queue of h-calls in the model with patient h-calls is larger than that in the model with impatient h-calls.

Another goal of performing numerical experiments was the estimation of the proposed formula accuracy. The exact values (EV) of the QoS metrics for the model with patient h-calls at the identical time of channel occupancy by different types of calls are determined by the following formulae which are easily derived from classical one-dimensional birth-and-death processes:

$$P_{\mathrm{o}} = 1 - \sum_{k=0}^{N-g-1} \rho_k,$$

$$\tilde{N} = \sum_{k=1}^{N-1} k\rho_k + N\left(1 - \sum_{k=0}^{N-1} \rho_k\right),$$

$$L_{\mathrm{h}} = \frac{A\tilde{\nu}_{\mathrm{h}}^{N+1}}{(1-\tilde{\nu}_{\mathrm{h}})^2},$$

where

$$\rho_k = \begin{cases} \dfrac{\nu^k}{k!}\cdot \rho_0 & \text{if } k = \overline{1, N-g}, \\ \left(\dfrac{\lambda}{\lambda_{\mathrm{h}}}\right)^{N-g} \cdot \dfrac{\nu_h^k}{k!} \cdot \rho_0 & \text{if } k = \overline{N-g+1, N}, \end{cases}$$

$$\rho_0 = \left(\sum_{k=0}^{N-g} \frac{\nu^k}{k!} + \left(\frac{\lambda}{\lambda_{\mathrm{h}}}\right)^{N-g} \sum_{k=N-g+1}^{N} \frac{\nu_h^k}{k!} + \left(\frac{\lambda}{\lambda_{\mathrm{h}}}\right)^{N-g} \cdot \frac{N^N}{N!} \cdot \frac{\tilde{\nu}_{\mathrm{h}}^{N+1}}{1-\tilde{\nu}_{\mathrm{h}}}\right)^{-1};$$

$$\lambda := \lambda_{\mathrm{o}} + \lambda_{\mathrm{h}}; \quad \nu := \frac{\lambda}{\mu}; \quad \mu := \mu_{\mathrm{o}} = \mu_{\mathrm{h}}; \ \tilde{\nu}_{\mathrm{h}} := \frac{\nu_{\mathrm{h}}}{N}; \ A := \left(\frac{\lambda}{\lambda_{\mathrm{h}}}\right)^{N-g} \cdot \frac{N^N}{N!}.$$

Note that approximate values (AV) of QoS metrics are almost identical to their exact values when the accepted assumption about ratios of load parameters of o- and h-calls is valid. Some comparisons for the models with parameters $N = 10$,

$\lambda_0 = 0.3$, $\lambda_h = 2$, $\mu_o = \mu_h = 3$ and $N = 15$, $\lambda_0 = 2$, $\lambda_h = 4$, $\mu_o = \mu_h = 5$ are presented in Tables 2.1 and 2.2, respectively.

As indicated in Tables 2.1 and 2.2 the accuracy of approximate formulae is sufficiently high even when the accepted assumption is not valid, i.e. $\mu_o = \mu_h$. Similar results are obtained for any possible values of initial data of the model.

It is worth noting that sufficiently high accuracy exists for the initial data not satisfying the above-mentioned assumption concerning the ratios of traffic of o- and h-calls. In other words, the numerical experiments for the models with symmetric traffic (i.e. $\lambda_o = \lambda_h$) in addition to the ones in which the intensity of o-calls greatly exceeds the intensity of h-calls showed a sufficiently high accuracy for the suggested approximate formulae. Certain results of the comparison for the models

Table 2.1 Comparison of exact and approximate values of QoS metrics for the model with patient h-calls in the case where $N = 10$, $\lambda_o = 0.3$, $\lambda_h = 2$, $\mu_o = \mu_h = 3$

	P_o		$\tilde{N}$		L_h	
g	EV	AV	EV	AV	EV	AV
1	1.25516E-07	1.26890E-07	0.766666654	0.76666665	1.28659E-09	8.27322E-10
2	1.48437E-06	1.50207E-06	0.766666518	0.76666652	1.11877E-09	8.27185E-10
3	1..56409E-05	1.58504E-05	0.766665103	0.76666508	9.72847E-10	8.26463E-10
4	1.44624E-04	1.46806E-04	0.766652204	0.76665199	8.45954E-10	8.23177E-10
5	1.15118E-03	1.17021E-03	0.766551548	0.76654965	7.35612E-10	8.01069E-10
6	7.68970E-03	7.81343E-03	0.765897696	0.76588532	6.39663E-10	7.72305E-10
7	4.16251E-02	4.20410E-02	0.762504156	0.76246247	5.56228E-10	6.80938E-10
8	1.73829E-01	1.72958E-01	0.749283793	0.74937084	4.83677E-10	5.22265E-10
9	5.21507E-01	5.11655E-01	0.714515980	0.71550113	4.20589E-10	3.33886E-10

Table 2.2 Comparison of exact and approximate values of QoS metrics for the model with patient h-calls in the case where $N = 15$, $\lambda_o = 2$, $\lambda_h = 4$, $\mu_o = \mu_h = 5$

	P_o		$\tilde{N}$		L_h	
g	EV	AV	EV	AV	EV	AV
1	4.68574E-11	4.8229E-11	1.200000000	1.20000000	4.67447E-13	3.54354E-13
2	5.48753E-10	5.65712E-10	1.200000000	1.20000000	3.11631E-13	3.54052E-13
3	5.97224E-09	6.17134E-09	1.199999998	1.20000000	2.07754E-13	3.53091E-13
4	6.00456E-08	6.22177E-08	1.199999976	1.19999998	1.38503E-13	3.50246E-13
5	5.53951E-07	5.75773E-07	1.199999778	1.19999977	9.23352E-14	3.42914E-13
6	4.65197E-06	4.85165E-06	1.199998139	1.19999806	6.15568E-14	3.26731E-13
7	3.52214E-05	3.68591E-05	1.199985911	1.19998526	4.10379E-14	2.96472E-13
8	2.37618E-04	2.49355E-04	1.199904953	1.19990026	2.73586E-14	2.49023E-13
9	1.407653E-03	1.478063E-03	1.199436939	1.19940877	1.82391E-14	1.8732E-13
10	7.18795E-03	7.51375E-03	1.19712482	1.1969945	1.21594E-14	1.21736E-13
11	3.090382E-02	3.186036E-02	1.187638471	1.18725586	8.10625E-15	6.58789E-14
12	1.087237E-01	1.091789E-01	1.156510523	1.15632842	5.40416E-15	2.87429E-14
13	3.032389E-01	2.945699E-01	1.078704409	1.08217203	3.60278E-15	9.97493E-15
14	6.476778E-01	6.191261E-01	0.940928864	0.95234956	2.40185E-15	2.82974E-15

with parameters $N = 10$, $\lambda_o = 2$, $\lambda_h = 0.3$, $\mu_o = \mu_h = 3$ and $N = 15$, $\lambda_o = \lambda_h = 4$, $\mu_o = \mu_h = 5$ are shown in Tables 2.3 and 2.4, respectively.

As seen from Tables 2.3 and 2.4 small deviations are observed upon calculation of average length of h-calls. Therewith the approximate values of the QoS metrics are always larger than their exact values. The last circumstance suggests that to increase the reliability of network performance the obtained approximate formulae can be applied at the initial stages of its design.

Analogous results are obtained for the models with a limited queue of h-calls. For the model with a limited queue of patient h-calls the approximate results obtained were compared with results in [11], wherein accurate formulae for the model with identical (with respect to channel occupation times) calls were developed. The

Table 2.3 Comparison of exact and approximate values of QoS metrics for the model with patient h-calls in the case where $N = 10$, $\lambda_o = 2$, $\lambda_h = 0.3$, $\mu_o = \mu_h = 3$

	P_o		$\tilde{N}$		L_h	
g	EV	AV	EV	AV	EV	AV
1	1.18332E-07	1.20335E-07	0.766666588	0.76666659	2.57292E-11	5.31368E-10
2	1.39066E-06	1.41037E-06	0.766665740	0.76666573	3.35598E-12	1.51235E-10
3	1.45316E-05	1.46932E-05	0.766656979	0.76665687	4.37736E-13	3.60410E-11
4	1.32920E-04	1.33885E-04	0.766578053	0.76657741	5.70961E-14	6.56823E-12
5	1.04287E-03	1.04528E-03	0.765971420	0.76596981	7.44731E-15	8.94899E-13
6	6.83047E-03	6.80289E-03	0.762113022	0.76213141	9.71388E-16	8.95273E-14
7	3.60318E-02	3.56001E-02	0.742645458	0.74293325	1.26703E-16	6.41398E-15
8	1.46785E-01	1.43779E-02	0.668809421	0.67081413	1.65265E-17	3.19136E-16
9	4.46385E-01	4.35614E-01	0.469076476	0.47625721	2.15562E-18	1.08186E-17

Table 2.4 Comparison of exact and approximate values of QoS metrics for the model with patient h-calls in the case where $N = 15$, $\lambda_o = \lambda_h = 4$, $\mu_o = \mu_h = 5$

	P_o		$\tilde{N}$		L_h	
g	EV	AV	EV	AV	EV	AV
1	1.76280E-09	1.86813E-09	1.599999999	1.6000000	2.62346E-11	2.73750E-11
2	1.54833E-08	1.64400E-08	1.599999988	1.59999999	1.31173E-11	2.65173E-11
3	1.26382E-07	1.34674E-07	1.599999899	1.59999989	6.55865E-12	2.51571E-11
4	9.52992E-07	1.01936E-06	1.599999238	1.59999918	3.27933E-12	2.28218E-11
5	6.59388E-06	7.07790E-06	1.599994725	1.59999434	1.63966E-12	1.93241E-11
6	4.15307E-05	4.46986E-05	1.599966775	1.59996424	8.19832E-13	1.48828E-11
7	2.35835E-04	2.54048E-04	1.599811332	1.59979676	4.09916E-13	1.01637E-11
8	1.19341E-03	1.28238E-03	1.599045275	1.59897409	2.04958E-13	6.00668E-12
9	5.30507E-03	5.65419E-03	1.595755945	1.59547665	1.02479E-13	3.00061E-12
10	2.03616E-02	2.13408E-02	1.583710735	1.58292732	5.12395E-14	1.23831E-12
11	6.61732E-02	6.74847E-02	1.547061432	1.54601217	2.56197E-14	4.13800E-13
12	1.78719E-01	1.75864E-01	1.457024740	1.45930802	1.28099E-14	1.10834E-13
13	3.95653E-01	3.76449E-01	1.283477649	1.29884051	6.40494E-15	2.40351E-14
14	7.10236E-01	6.69481E-01	1.031811366	1.06441553	3.20247E-15	4.38201E-15

comparative analysis was performed over a wide range of variance in the initial data of the model. It is worth noting that the approximate and exact results from calculating the loss probability of h-calls were identical. An insignificant difference is observed in calculating the blocking probability of o-calls. It can be seen that the maximal value of error in all executed experiments was negligible and in most cases equals zero even for data not satisfying the above-mentioned assumption concerning the ratios of traffic of o- and h-calls. So, for example, at $N=50$, $B=10$, $\lambda_o=2$, $\lambda_h=1$, $\mu=10$ the maximal error holds for $g=49$ and is 1.6%, i.e. for these initial data the exact value of P_o equals 0.233 while its approximate value equals 0.249.

2.2 Models with Queues for o-Calls

In order to compensate for the chance of o-calls a queue (limited or unlimited) is required for them, while maintaining a high chance for h-calls to access the system via reservation of channels. Here we consider four types of models with queues for o-calls whereas h-calls are treated according to a lost model [2, 9]. Note that in this section it is assumed that the required handling time is independent of call type and exponentially distributed with the same average μ^{-1}.

In all schemes it is assumed that all $m+n$ channels are divided into two groups: a Primary group with m channels and a Secondary group with n channels. If all channels in both groups are busy the h-call is lost. New calls are only allowed to the Primary group; therefore if all m channels are busy this call is placed into a queue.

In two schemes reallocation of channels from one group to another is not allowed, i.e. isolated reservation is considered. In the HOPS (Handoff calls Overflow from Primary to Secondary) scheme, for handling of an h-call, the Primary group channels are used first and upon absence of an empty channel in the Primary group (all m channels are busy) the Secondary group channels are used. In the HOSP (Handoff calls Overflow from Secondary to Primary) scheme the search for a free channel for handling of an h-call is realized in the Secondary group first.

The last two schemes involve shared reservation of channels. They maybe described as follows. In one scheme upon release of a channel in the Primary group (either by an o-call or h-call) an h-call from the Secondary group is reallocated to the Primary group regardless of the length of the o-calls in the queue. However, o-calls from the queue are admitted to channels in accordance with the FIFO discipline only if the number of total empty channels exceeds n. The given channel reallocation scheme is called Handoff Reserve Margin Algorithm (HRMA). The main difference of the final scheme from the previous one is in the reallocation of the h-call from the Secondary group to the Primary group. Upon release of a channel in the Primary group (either by an o-call or h-call) an h-call from the Secondary group is reallocated to the Primary group if and only if there are no o-calls in the queue. In other words reallocation of h-calls from the Secondary group to the Primary group is not allowed until the queue does not contain any o-calls. This reservation scheme is called Handoff calls Overflow from Primary to Secondary with Rearrangement (HOPSWR).

2.2.1 Models Without Reassignment of Channels

Here we consider two isolated reservation schemes in which reallocation (reassignment) of channels from one group to another is not allowed. In the HOPS (Handoff calls Overflow from Primary to Secondary) scheme the initial search for a free channel for service of an h-call is carried out in the Primary Group and upon the absence of an empty channel in the Primary group (all m channels are busy) Secondary group channels are used (see Fig. 2.8a). In the HOSP (Handoff calls Overflow from Secondary to Primary) scheme the initial search for a free channel to service an h-call is carried out in the Secondary Group (see Fig. 2.8b).

In both schemes o-calls can be served only in the Primary Group of channels and any conservative discipline of service which does not admit to idle times of channels in the presence of a queue can be used for service of the o-calls queue. In both schemes in cases of occupancy of all $m+n$ channels the h-call is lost and any reassignment of the h-call from one group to another is not permitted.

First of all we shall consider the HOPS scheme in the model with an infinite queue of o-calls. The QoS metrics include probability of loss of h-calls (P_{h}), average length of o-calls queue (L_{o}), and the average latency period in the queue (W_{o}).

The following hierarchical approach can be used for the analysis of this model as in this scheme received h-calls initially go to the Primary Group of channels, and

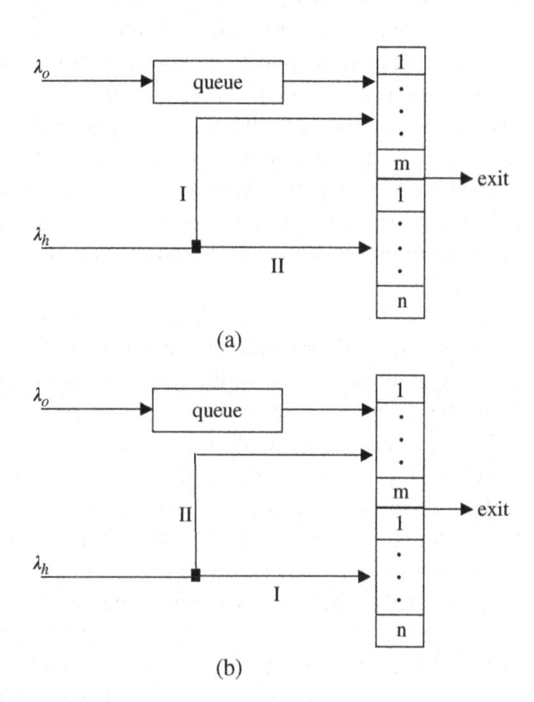

Fig. 2.8 Diagram of the HOPS (**a**) and HOSP (**b**) model

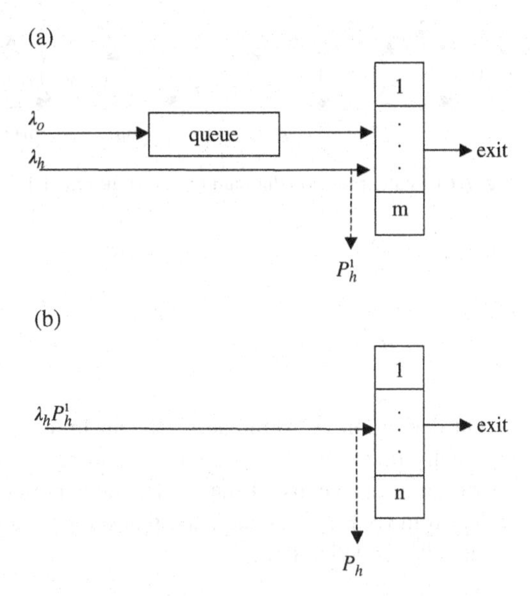

Fig. 2.9 Hierarchical approach for study of the HOPS model

only missed calls of the given type are further received in the Secondary Group of channels. In the first step of the hierarchy (see Fig. 2.9a) we shall consider a system with m channels which serve calls of two types with rates λ_o and λ_h, thus the holding time of any type of call has an exponential distribution with a common mean μ^{-1}. New calls are buffered in an infinite queue and h-calls are lost in case of occupancy of all channels. Missed h-calls are forwarded to the Secondary Group of channels for service.

Consider that the probability of loss of h-calls is equal to P_h^1 in the Primary Group of channels. From Poisson flow property we deduce that the input to the Secondary Group of channels forms Poisson flow with rate $\tilde{\lambda}_h := \lambda_h P_h^1$. Hence on the second stage of hierarchy (see Fig. 2.9b) we examine classic Erlang model *M/M/n/0*. Loss probability of calls of this model then will be considered as the desired P_h. And the desired L_o and W_o are calculated as QoS parameters of queuing system described at the first step of hierarchy.

Now we shall consider the problem of calculation of the above-specified QoS metrics. The state of the queuing system with two types of calls described in the first step of the hierarchy is given by scalar parameter k which specifies the total number of calls in the system, $k = 0,1,2,\ldots$. Stationary distribution of the corresponding one-dimensional birth-death process (1-D BDP) is calculated as (Fig. 2.10):

$$\rho_k = \begin{cases} \dfrac{v^k}{k!}\rho_0 & \text{if } 1 \le k \le m, \\ \dfrac{v^m}{m!} \cdot \dfrac{v_o^{k-m}}{m^{k-m}}\rho_0 & \text{if } k \ge m+1, \end{cases} \tag{2.35}$$

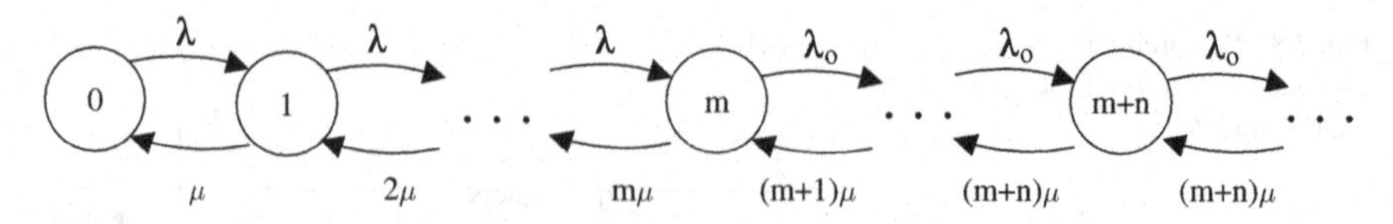

Fig. 2.10 State transition diagram for the HOPS model

where

$$\nu := \lambda/\mu,\ \lambda := \lambda_o + \lambda_h,\ \nu_o := \lambda_o/\mu,\ \rho_0 = \left(\sum_{k=0}^{m} \frac{\nu^k}{k!} + \frac{\nu^m}{m!} \cdot \frac{\nu_o}{m - \nu_o}\right)^{-1}.$$

On derivation of formulae (2.35) an intuitively clear and simple condition of ergodicity of the models becomes apparent: $\nu_o < m$. Hence we can see that the condition of ergodicity of the models does not depend on the loading of handover calls. From (2.35) it is concluded that the probability of h-call loss in the Primary Group (P_h^1) is defined as:

$$P_h^1 = 1 - \sum_{k=0}^{m-1} \rho_k . \tag{2.36}$$

Hence, required QoS metric P_h is calculated by means of Erlang's classical B-formula for the *M/M/n/0* c model with a load of $\tilde{\nu}_h := \tilde{\lambda}_h/\mu$ Erl. In other words, $P_h = E_B(\tilde{\nu}_h, n)$.

After certain transformations we obtain the following formula for calculation of QoS metric L_o:

$$L_o = \sum_{k=1}^{\infty} k\rho_{k+m} = \frac{\nu^m}{(m-1)!} \cdot \frac{\nu_o}{(m - \nu_o)^2} \cdot \rho_0 . \tag{2.37}$$

QoS metric W_o is obtained from the Little formulae, i.e. $W_o = L_o/\lambda$.

The developed approach allows the definition of QoS metrics of the model also in the presence of a limited buffer for waiting in a queue of o-calls. We should note that in these models at any loading and structural parameter values in the system there is a stationary mode, i.e. ergodicity performance is not required, $\nu_o < m$.

Let the maximum size of the buffer be equal to R, $R<\infty$. On the basis of formulae for finite BDP we conclude that stationary distribution of the appropriate system is calculated as follows:

$$\rho_k = \begin{cases} \dfrac{\nu^k}{k!}\rho_0 & \text{если} \quad 1 \le k \le m, \\ \dfrac{\nu^m}{m!} \cdot \dfrac{\nu_o^{k-m}}{m^{k-m}}\rho_0 & \text{если} \quad m+1 \le k \le m+R, \end{cases} \tag{2.38}$$

where

$$\rho_0 = \left(\sum_{k=0}^{m} \frac{\nu^k}{k!} + \frac{\nu^m}{m!} \cdot \frac{m^m}{\nu_o^m} \sum_{k=m+1}^{R} \left(\frac{\nu_o}{m} \right)^k \right)^{-1} .$$

Then by using (2.38) from (2.36) and Erlang's B-formula QoS metric P_h is calculated. And from (2.38) the following formula for calculation of the QoS metric $L_o(R)$ for the model with a limited queue of o-calls finds:

$$L_o(R) = \sum_{k=1}^{R} k \rho_{k+m} . \quad (2.39)$$

From (2.39) quantity $W_o(R)$ in the given model is calculated as follows

$$W_o(R) = \frac{L_o(R)}{\lambda_o (1 - P_o(R))} ,$$

where $P_o(R)$ denotes the loss probability of o-calls in the given model, i.e. $P_o(R) = \rho_{m+R}$.

Now consider the HOSP scheme in the model with an infinite queue of o-calls. As in the previous scheme it is possible to use the hierarchical approach. Here in the first step of the hierarchy (see Fig. 2.11a) the classical Erlang model *M/M/n/0* with

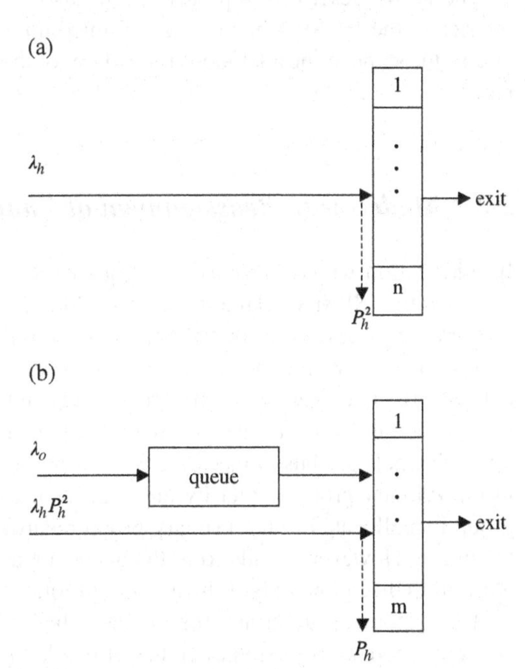

Fig. 2.11 Hierarchical approach for study of the HOSP model

a load $\nu_h := \lambda_h/\mu$ Erl is considered. The probability of h-call loss in this model will be denoted by P_h^2, i.e. $P_h^2 = E_B(\nu_h, n)$.

Missed h-calls in this system are forwarded to the Primary Group for reception of service. Hence, in the second step of the hierarchy (see Fig. 2.11b) the system with m channels which serves calls of two types with rates λ_o and $\hat{\lambda}_h := \lambda_h P_h^2$, thus the holding time of a call of any type that has an exponential distribution with the general average μ^{-1}, is considered. New calls are buffered in an infinite queue and in the case of occupancy of all channels h-calls are lost. Missed h-calls in this system are finally lost. Thus, stationary distributions of the queuing system described in the second step of the hierarchy are calculated by means of the formula (2.35). However, in this case the specified formula parameter λ is determined as $\lambda = \lambda_o + \hat{\lambda}$.

Stationary distribution of the last system will be denoted through σ_k, $k = 0,1,2,\ldots$. The condition of ergodicity of models in the given scheme is also $v_o < m$. Then in view of the above-stated, it is concluded that the required QoS metric P_h in the HOSP scheme of channel distribution is calculated as:

$$P_h = 1 - \sum_{k=0}^{m-1} \sigma_k . \tag{2.40}$$

Other QoS metrics L_o and W_o in the given scheme of channel allocation are also calculated from (2.37) and Little's formula, accordingly. Thus, it is necessary to consider that in this case $\lambda = \lambda_o + \hat{\lambda}$.

The above-described approach might also be used for calculation of the QoS metrics in the HOSP scheme with a limited queue of o-calls. As this procedure is almost the same as the analogous procedure for the HOPS scheme, it is not presented here.

2.2.2 Models with Reassignment of Channels

In this section we consider two schemes with reassignment of channels. First we consider the HRMA (Handoff Reserve Margin Algorithm) scheme for channel assignment. In this case for the handling of an h-call a channel from the Primary group is used first and upon absence of an empty channel in this group (all m channels are busy) Secondary group channels are used. If all channels in both groups are busy the h-call is lost. New calls are only allowed to the Primary group; therefore if all m channels are busy this call is placed into a queue. Upon release of a channel in the Primary group (either by an o-call or h-call) an h-call from the Secondary group is reallocated to the Primary group regardless of the length of the o-calls in the queue. However, o-calls from the queue are admitted to channels in accordance with FIFO discipline only if the number of total empty channels exceeds n.

The system's state at any time is described by the two-dimensional vector $\mathbf{k} = (k_1, k_2)$, where k_1 is the total number of o-calls in the system $k_1 = 0,1,2,\ldots$, and k_2

is the total number of busy channels, $k_2=0,1,\ldots, m+n$. Note that state space S of the given model does not include vectors $\mathbf{k}$ with components $k_1>0$, $k_2<m$.

On the basis of the adopted channel allocation scheme, we can conclude that elements of the generating matrix $q(\mathbf{k}, \mathbf{k}')$, $\mathbf{k}, \mathbf{k}' \in S$ of the appropriate 2-D MC are defined from the following relations (see Fig. 2.12):

$$q(\mathbf{k},\mathbf{k}')=\begin{cases}\lambda_o+\lambda_h, & \text{if } k_1=0,\ k_2\le m-1,\ \mathbf{k}'=\mathbf{k}+\mathbf{e}_2,\\ \lambda_h, & \text{if } k_2\ge m,\ \mathbf{k}'=\mathbf{k}+\mathbf{e}_2,\\ \lambda_o, & \text{if } k_2\ge m,\ \mathbf{k}'=\mathbf{k}+\mathbf{e}_1,\\ k_2\mu, & \text{if } k_2\ne m,\ \mathbf{k}'=\mathbf{k}-\mathbf{e}_2,\\ m\mu, & \text{if } k_2=m,\ \mathbf{k}'=\mathbf{k}-\mathbf{e}_1,\\ 0 & \text{in other cases.}\end{cases} \tag{2.41}$$

The desired QoS metrics for the model with an infinite queue of o-calls are calculated by stationary distribution of the model as follows:

$$P_h=\sum_{k_1=0}^{\infty}p(k_1,m+n), \tag{2.42}$$

$$L_o=\sum_{k_1=1}^{\infty}\sum_{k_2=m}^{m+n}k_1p(k_1,k_2), \tag{2.43}$$

Here we develop simple approximate computational procedures to calculate these QoS metrics. For correct application of the approximate approach condition

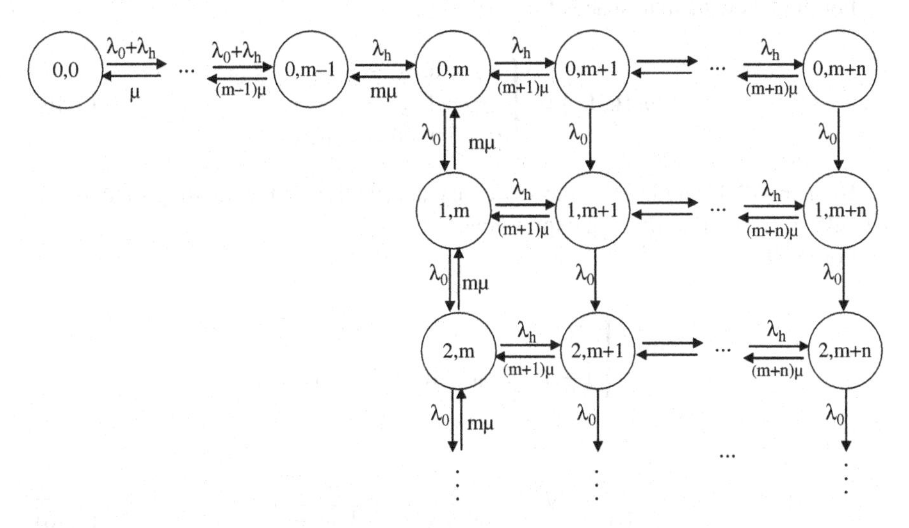

Fig. 2.12 State transition diagram for the HRMA model with an unlimited queue of patient o-calls

$\lambda_h >> \lambda_o$is required. As was mentioned in previous sections this condition is true for micro- and picocells. In addition, we assume that both types of calls are characterized by very short duration with respect to their frequency of arrivals. The last assumption is valid for many real system models that are quite close to those in this work (see Sect. 1.1).

The following splitting of state space of the given model is considered:

$$S = \bigcup_{i=0}^{\infty} S_i, S_i \bigcap S_j = \varnothing, i \neq j, \tag{2.44}$$

where

$$S_i := \{k \in S : k_1 = i\}.$$

Furthermore, the class of microstates S_i is merged into the isolated merged state $<i>$ and an appropriate merged model with state space $\tilde{S} := \{< i >: i = 0, 1, 2, \ldots\}$ is constructed.

The elements of the generated matrix of splitting models with state space S_i that is denoted by $q_i(\mathbf{k}, \mathbf{k}')$, $\mathbf{k}, \mathbf{k}' \in S_i$ are calculated as follows [see (2.41)]:

For the model with state space S_0:

$$q_0\left(\mathbf{k}, \mathbf{k}'\right) = \begin{cases} \lambda_o + \lambda_h, & \text{if } k_2 \leq m-1, \quad k_2{}' = k_2 + 1, \\ \lambda_h, & \text{if } m \leq k_2 \leq m+n-1, \quad k_2{}' = k_2 + 1, \\ k_2\mu, & \text{if } k_2{}' = k_2 - 1, \\ 0 & \text{in other cases,} \end{cases} \tag{2.45}$$

For models with state spaces S_i, $i \geq 1$:

$$q_i\left(\mathbf{k}, \mathbf{k}'\right) = \begin{cases} \lambda_h & \text{if } k_2{}' = k_2 + 1, \\ k_2\mu & \text{if } k_2{}' = k_2 - 1, \\ 0 & \text{in other cases.} \end{cases} \tag{2.46}$$

By using (2.45) and (2.46) the stationary distribution of the splitting models are calculated from the following expressions:

For $i = 0$:

$$\rho^0(j) = \begin{cases} \frac{\nu^j}{j!}\rho_0 & \text{if } j = \overline{1, m}, \\ \left(\frac{\nu}{\nu_h}\right)^m \frac{\nu_h{}^j}{j!}\rho_0 & \text{if } j = \overline{m+1, m+n}. \end{cases} \tag{2.47}$$

For $i>0$:

$$\rho^i(j) = \frac{m!}{\nu_h{}^m} \frac{\nu_h{}^j}{j!}\rho_1, \; j = \overline{m+1, m+n}, \tag{2.48}$$

where

$$\rho_0 = \left(\sum_{i=0}^{m} \frac{v^i}{i!} + \left(\frac{v}{v_h}\right)^m \sum_{i=m+1}^{m+n} \frac{v_h^i}{i!} \right)^{-1}, \quad \rho_1 = \left(\frac{m!}{{v_h}^m} \sum_{i=m}^{m+n} \frac{v_h^i}{i!} \right)^{-1}, v := v_o + v_h.$$

By using (2.45), (2.46), (2.47), and (2.48) we conclude that the elements of the generating matrix of merged model $q\left(< i >, < i' >\right)$, $< i >, < i' > \in \tilde{S}$ are calculated as follows:

$$q\left(< i >, < i' >\right) = \begin{cases} \lambda_o \sum_{j=m}^{m+n} \rho^0(j) & \text{if } i = 0,\ i' = i + 1, \\ \lambda_o & \text{if } i \geq 0,\ i' = i + 1, \\ m\mu\rho_1 & \text{if } i \geq 0,\ i' = i - 1, \\ 0 & \text{in other cases.} \end{cases} \tag{2.49}$$

From (2.49) we find the following ergodicity condition of the merged model:

$$a := \frac{v_o}{m\rho^1(m)} < 1$$

or in explicit form

$$\frac{v_o}{m} \cdot \frac{m!}{v_h^m} \left(\sum_{i=m}^{m+n} \frac{v_h^i}{i!} \right) < 1. \tag{2.50}$$

Note 2.3. It is important to note that ergodicity condition (2.50) is exactly the stability condition of the system established in [2], i.e. here we easily obtain the stability condition of the investigated model.

By fulfilling the condition (2.50) the stationary distribution of the merged model $(\pi(< i >) :< i > \in \tilde{S})$ is calculated as

$$\pi(< i >) = a^i b \pi(< 0 >), \quad i = 1, 2, \ldots, \tag{2.51}$$

where

$$b := \sum_{i=m}^{m+n} \rho^0(i), \quad \pi(< 0 >) = \frac{1 - a}{1 - a + ab}.$$

In summary the following simple formulae for calculation of the desired QoS metrics (2.42) and (2.43) can be suggested:

$$P_h \approx \frac{1}{1 - a + ab}((1 - a)E_B(v_h, m + n) + abE_B^T(v_h, m + n)), \tag{2.52}$$

$$L_o \approx \frac{ab}{(1 - a + b)(1 - a)}, \tag{2.53}$$

where $E_B^T(v_h, m+n)$ – the truncated Erlang's B-formula, i.e.

$$E_B{}^T(v_h, m+n) := \frac{v_h{}^{m+n}}{(m+n)!}\left(\sum_{i=m}^{m+n} \frac{v_h{}^i}{i!}\right)^{-1}.$$

From (2.52) we conclude that P_h is the convex combination of two functions $E_B(v_h, m+n)$ and $E_B^T(v_h, m+n)$. In other words, for any admissible values of number of channels and traffic loads the following limits for P_h may be proposed:

$$E_B(v_h, m+n) \le P_h \le E_B^T(v_h, m+n). \tag{2.54}$$

The limits in (2.54) will be achieved and are the same only in the special case $m=0$, i.e. when only h-calls arrive in the system.

The proposed approach also allows calculation of QoS metrics for the model with a limited queue of o-calls. Indeed, let the maximal length of the queue of o-calls be R, $R<\infty$. Then for any admissible values of number of channels and traffic loads in this system the stationary mode exists, i.e. in this case fulfilling of the ergodicity condition (2.50) is not required.

For the given model the number of splitting models is $R+1$ and their stationary distribution are calculated by (2.47) and (2.48). Making use of the above-described approach and omitting the known transformation the following expressions are determined to calculate the stationary distribution of the merged model:

$$\pi(<i>) = a^i b\pi(<0>),\ i = \overline{1,R}, \tag{2.55}$$

where

$$\pi(<0>) = \left(1 + ab\frac{1-a^R}{1-a}\right)^{-1}.$$

Since the approximate values of the QoS metrics for the model with a limited queue of o-calls are calculated as follows:

$$P_h(R) \approx \pi(<0>)E_B(v_h, m+n) + (1-\pi(<0>))E_B^T(v_h, m), \tag{2.56}$$

$$L_o(R) \approx ab\frac{1-a^R(R+1+Ra)}{(1-a)^2}\pi(<0>), \tag{2.57}$$

$$W_o(R) \approx \frac{L_o(R)}{\lambda_o(1-P_o(R))}, \tag{2.58}$$

where $P_o(R)$ is the probability of loss of o-calls which is calculated by

$$P_o(R) \approx \pi(R) or P_o(R) \approx a^R b\pi(<0>). \tag{2.59}$$

Now we consider another shared reservation scheme of channels for h-calls in the model with a queue of o-calls. The main difference between this scheme and the previous one is reallocation of h-calls from the Secondary group to the Primary group. Upon release of a channel in the Primary group (either by an o-call or h-call) an h-call from the Secondary group is reallocated to the Primary group if and only if there are no o-calls in the queue. In other words reallocation of h-calls from the Secondary group to the Primary group is not allowed until the queue does not contain any o-calls. This reservation scheme is called Handoff calls Overflow from Primary to Secondary with Rearrangement (HOPSWR).

The system's state at any time is described by the two-dimensional vector $\mathbf{k} = (k_1, k_2)$, where k_1 is the total number of busy channels $k_1 = 0,1,\ldots,m+n$, and k_2 is the number of o-calls in the queue, $k_2 = 0,1,2,\ldots$. The model's state space S has the following view:

$$S = \bigcup_{i=0}^{n} S_i\,, S_i \bigcap S_j = \emptyset,\ i \neq j, \tag{2.60}$$

where

$$\begin{aligned} S_0 &= \{(j,0) : j = 0,1,\ldots,m\} \cup \{(m,j) : j = 1,2,\ldots\}, \\ S_i &= \{(m+i,j) : j = 0,1,2,\ldots\},\ i \geq 0. \end{aligned}$$

On the basis of the adopted channel allocation scheme, we can conclude, that elements of the generating matrix $q(\mathbf{k},\mathbf{k}')$, $\mathbf{k},\mathbf{k}' \in S$ of appropriate 2-D MC are defined from the following relations (see Fig. 2.13):

$$q(\mathbf{k},\mathbf{k}') = \begin{cases} \lambda_o + \lambda_h & \text{if } k_1 \leq m-1,\ \mathbf{k}' = \mathbf{k} + \mathbf{e}_1, \\ \lambda_h & \text{if } k_1 \geq m,\ \mathbf{k}' = \mathbf{k} + \mathbf{e}_1, \\ \lambda_o & \text{if } k_1 \geq m,\ \mathbf{k}' = \mathbf{k} + \mathbf{e}_2, \\ k_1\mu & \text{if } k_1 \leq m-1,\ \mathbf{k}' = \mathbf{k} - \mathbf{e}_2, \\ m\mu & \text{if } k_1 \geq m,\ \mathbf{k}' = \mathbf{k} - \mathbf{e}_2, \\ k_1\mu & \text{if } k_1 \geq m,\ k_2 = 0,\ \mathbf{k}' = \mathbf{k} - \mathbf{e}_1, \\ (k_1 - m)\mu & \text{if } k_1 \geq m,\ k_2 \geq 0,\ \mathbf{k}' = \mathbf{k} - \mathbf{e}_1, \\ 0 & \text{in other cases.} \end{cases} \tag{2.61}$$

The desired QoS metrics of the system in this case are defined via stationary distribution of the model as follows:

$$P_h = \sum_{i=0}^{\infty} p(m+n,i), \tag{2.62}$$

$$L_o = \sum_{i=1}^{\infty} ip(i), \tag{2.63}$$

where $p(i) := \sum_{k \in S} p(\mathbf{k})\delta(k_2,i)$ are marginal probability mass functions.

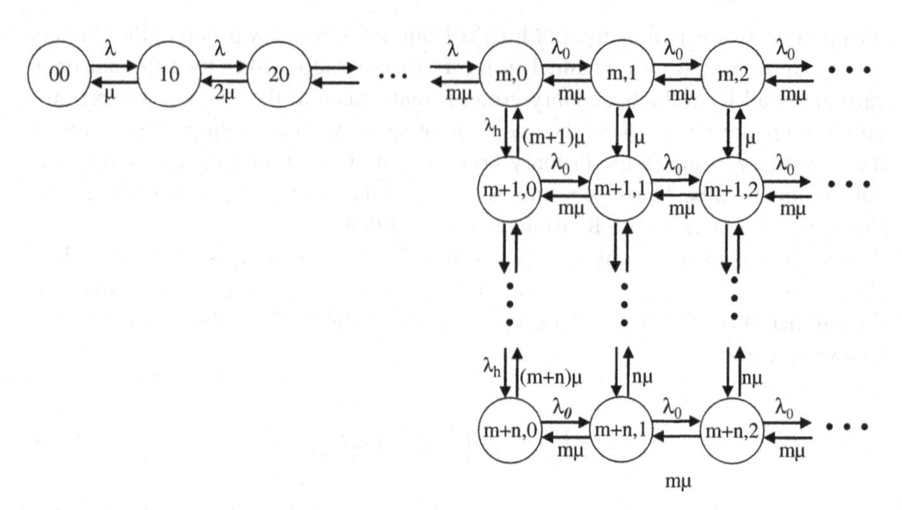

Fig. 2.13 State transition diagram for the HOPSWR model with an unlimited queue of patient o-calls

First, we will analyze the model of a macrocell with an infinite queue for o-calls. It is clear that the following condition holds true in macrocells $\lambda_o >> \lambda_h$.

The above-mentioned condition on relations of intensities of different types of calls allow the conclusion that transitions from state $\mathbf{k} \in S$ into state $\mathbf{k} + \mathbf{e_2} \in S$ occur more often than into state $\mathbf{k} + \mathbf{e_1} \in S$. In other words transition between states (microstates) inside classes S_i occurs more often than transitions between states from different classes. Because of this fact, classes of microstates S_i in (2.60) are depicted as isolated merged state <*i*>, and in initial state space S a known merging function is introduced. Thus, an appropriate 1-D MC with state space $\tilde{S} := \{< i >: i = 0, 1, 2, \ldots, n\}$ is constructed.

Elements of the generating matrix of split models with state space S_i, denoted as $q_i(\mathbf{k}, \mathbf{k}')$, $\mathbf{k}, \mathbf{k}' \in S_i$, are found through relations (2.61):

For the model with state space S_0:

$$q_0\left(\mathbf{k}, \mathbf{k}'\right) = \begin{cases} \lambda_o + \lambda_h & \text{if } k_1 \leq m-1,\ \mathbf{k}' = \mathbf{k} + \mathbf{e_1}, \\ \lambda_o & \text{if } k_1 \geq m,\ \mathbf{k}' = \mathbf{k} + \mathbf{e_2}, \\ k_1\mu & \text{if } k_1 \leq m-1,\ \mathbf{k}' = \mathbf{k} - \mathbf{e_1}, \\ m\mu & \text{if } k_1 \geq m,\ \mathbf{k}' = \mathbf{k} - \mathbf{e_2}, \\ 0 & \text{in other cases;} \end{cases} \tag{2.64}$$

For models with state space S_i, $i \geq 1$:

$$q_i\left(\mathbf{k}, \mathbf{k}'\right) = \begin{cases} \lambda_o, & \text{if } \mathbf{k}' = \mathbf{k} + \mathbf{e_2}, \\ m\mu, & \text{if } \mathbf{k}' = \mathbf{k} - \mathbf{e_2}, \\ 0 & \text{in other cases.} \end{cases} \tag{2.65}$$

The stationary probability of state $\mathbf{k} \in S$ inside the split model with state space S_i is denoted $\rho(\mathbf{k})$. Then with the aid of (2.64) and (2.65) the stationary distribution of split models can be found:

For the model with state space S_0:

$$\rho(i,j) = \begin{cases} \frac{v^i}{i!}\rho_0 & \text{if } 1 \le i \le m,\ j = 0, \\ \frac{v^m}{m!}\tilde{v}_o{}^j \rho_0 & \text{if } i = m,\ j \ge 1, \end{cases} \quad (2.66)$$

where

$$\rho_0 = \left(\sum_{i=0}^{m} \frac{v^i}{i!} + \frac{v^m}{m!} \cdot \frac{\tilde{v}_o}{1 - \tilde{v}_o} \right)^{-1}, \ v = v_o + v_h,\ \tilde{v}_o = v_o/m;$$

For models with state space $S_i,\ i = 1, 2, \dots, n$:

$$\rho\,(m + i, j) = \tilde{v}_o{}^j\,(1 - \tilde{v}_o)\,,\ j = 0, 1, 2, \dots\ . \quad (2.67)$$

Upon derivation of formula (2.67) we obtain an intuitively clear and simple ergodicity condition of the model: $\tilde{v}_o < 1$. It is seen that this condition does not depend on h-call load.

Transition intensities between merged states $< i >,\ < i' > \in \tilde{S}$ that are denoted $q\left(< i >, < i' >\right) \in \tilde{S}$ are found with the aid of (2.61), (2.66), and (2.67), i.e.

$$q\left(< i >,\ < i' >\right) = \begin{cases} \lambda_h c & \text{if } i = 0,\ i' = 1, \\ \lambda_h & \text{if } 1 \le i \le n - 1,\ i' = i + 1, \\ (m + i)\,(1 - \tilde{v}_o) + i\mu\tilde{v}_o & \text{if } 1 \le i \le n,\ i' = i - 1, \\ 0 & \text{in other cases} \end{cases} \quad (2.68)$$

where

$$c := 1 - \rho_0 \sum_{j=0}^{m-1} \frac{v^j}{j!} \ .$$

With the ergodicity condition of the system holding true, from relation (2.68) the stationary distribution of a merged model is defined as follows:

$$\pi(< i >) = c \prod_{j=1}^{i} \frac{\lambda_h{}^j}{(m + j)\,(1 - \tilde{v}_o) + j\mu\tilde{v}_o} \pi(< 0 >), \quad i = 1, 2, \dots, n, \quad (2.69)$$

where

$$\pi\left(<0>\right)=\left(1+c\sum_{i=1}^{n}\prod_{j=1}^{i}\frac{\lambda_{h}{}^{j}}{(m+j)\left(1-\tilde{v}_{o}\right)+j\mu\tilde{v}_{o}}\right)^{-1}.$$

After required mathematical transformations we obtain the following approximate formulae for calculating the QoS metrics of the initial model:

$$P_{h}\approx\pi\left(<n>\right)\,, \tag{2.70}$$

$$L_{o}\approx\frac{\tilde{v}_{o}}{1-\tilde{v}_{o}}\,, \tag{2.71}$$

Note 2.4. From formula (2.71) it can be seen that the average queue length of o-calls and appropriate average queue waiting time does not depend on the load of h-calls. This has a simple explanation in macrocells, since in such cells the intensity of o-calls is much higher than the intensity of h-calls, hence the load of h-calls on the Primary Group of channels is negligible.

The supposed method allows calculation of QoS metrics of a macrocell with a finite buffer for o-calls as well. Let $R, R<\infty$ be the maximum allowable length of the queue for o-calls. Then at any structural and load parameter values there is a stationary regime in the system, which has no need for ergodicity condition $\tilde{v}_{o}<1$ to be true.

In this case the state space of the initial model $S(R)$ is defined as:

$$S(R)=\bigcup_{i=0}^{n}S_{i}(R)\,,S_{i}(R)\cap S_{j}(R)=\emptyset,\ i\neq j, \tag{2.72}$$

where

$$\begin{aligned}S_{0}(R)&=\{(j,0):j=0,1,\ldots,m\}\cup\{(m,j):j=1,2,\ldots,R\}\,,\\ S_{i}(R)&=\ \{(m+i,j):j=0,1,2,\ldots,R\}\,.\end{aligned}$$

Using the above-described method and dropping in-between mathematical transformations, we determine that for the given model the stationary distribution of the merged model is defined as:

$$\pi_{R}(<i>)=c\prod_{j=1}^{i}\frac{\lambda_{h}{}^{j}}{(m+j)\,d+j\mu\,(1-d)}\,\pi_{R}(<0>),\ i=1,2,\ldots,n, \tag{2.73}$$

where

$$\pi_R(<0>) = \left(1 + c\sum_{i=1}^{n}\prod_{j=1}^{i}\frac{\lambda_h^j}{(m+j)\,d + j\mu\,(1-d)}\right)^{-1},\ d := \frac{1-\tilde{\nu}_o}{1-\tilde{\nu}_o^{R+1}}\ .$$

Consequently, estimated QoS values for the model with a finite queue are calculated as follows:

$$P_h(R) \approx \pi_R(<n>), \tag{2.74}$$

$$L_o(R) \approx d\sum_{i=1}^{R} i\tilde{\nu}_o^i\,, \tag{2.75}$$

$$W_o(R) \approx \frac{L_o(R)}{\lambda_o\,(1 - P_o(R))}\,, \tag{2.76}$$

where $P_o(R)$ is the loss probability of o-calls, which, for this model, is calculated as shown below:

$$P_o(R) \approx \tilde{\nu}_o^R\left(\frac{\nu^m}{m!}\pi_R(<0>) + d\,(1 - \pi_R(<0>))\right)\ . \tag{2.77}$$

Note 2.5. From formula (2.75) it can be seen that the average queue length of o-calls for the model with a finite queue also does not depend on h-call load. This is obvious in macrocells. However, in this model average queue waiting of o-calls depends on h-call load [see formula (2.76)]. But on the other hand, this dependency occurs only at small values of buffer for o-calls and disappears with growing buffer size.

Now we will consider calculation of QoS metrics for micro-cell models. As was noted above the following condition $\lambda_o << \lambda_h$ holds true in microcells. This means that transition from state $\mathbf{k} \in S$ into state $\mathbf{k} + \mathbf{e_1} \in S$ occurs more often than into state $\mathbf{k} + \mathbf{e_2} \in S$. In this case the following splitting of state space S is studied:

$$S = \bigcup_{i=0}^{\infty}\tilde{S}_i\,,\ \tilde{S}_i \bigcap \tilde{S}_{i'} = \varnothing,\ i \neq i'\,, \tag{2.78}$$

where

$$\tilde{S}_0 = \{(j,0) : j = 0,1,\ldots,m+n\}\,,\ \tilde{S}_i = \{(j,i) : j = m, m+1,\ldots,m+n\}\,, i \geq 1.$$

According to the above-mentioned condition on relations of intensities for different types of calls, in (2.78) transitions between microstates inside $\tilde{S}_i$ classes occur more often than between states from different classes.

Since the selected scheme of splitting of the initial state space completely defines the structure of the split and merged models, then further procedures for approximate calculation of the stationary distribution of the initial model are obvious. That is why below we drop some known interim steps for solution of this problem.

The stationary distribution of the split model with state space $\tilde{S}_0$ coincides with the appropriate stationary distribution of the Erlang model *M/M/m+n/0* with state-dependent rate $\lambda(j)$, i.e.

$$\lambda(j) = \begin{cases} \lambda_o + \lambda_h & \text{if } j \le m \\ \lambda_h & \text{if } j \ge m. \end{cases}$$

Thus, the stationary distribution of the split model with state space $\tilde{S}_0$ is calculated as follows:

$$\rho_0(j) = \begin{cases} \dfrac{\nu^j}{j!}\rho_0(0) & \text{if } j = 1, \ldots, m, \\ \left(\dfrac{\nu}{\nu_h}\right)^m \cdot \dfrac{\nu_h{}^j}{j!}\rho_0(0) & \text{if } j = m+1, \ldots, m+n, \end{cases} \tag{2.79}$$

where

$$\rho_0(0) = \left(\sum_{j=0}^{m}\frac{\nu^j}{j!} + \left(\frac{\nu}{\nu_h}\right)^m \sum_{j=m+1}^{m+n}\frac{\nu_h{}^j}{j!}\right)^{-1}.$$

Stationary distributions of split models with state spaces $\tilde{S}_i$, $i \ge 1$, are equal and coincide with the appropriate distribution of the classical Erlang model *M/M/n/0* with load ν_h.

Since the number of micro-state classes in (2.78) is infinite, then in this case the merged model represents 1-D MC with infinite state space $S' = \{<i>: i = 0, 1, 2, \ldots\}$. Here merged state $<i>$ comprises all microstates from class $\tilde{S}_i$. Then, considering (2.61) and the above-mentioned facts about stationary distributions inside split models, elements of the generating matrix of this given merged model are found:

$$q(<i>, <i'>) = \begin{cases} \lambda_o f & \text{if } i = 0,\ i' = 1 \\ \lambda_o & \text{if } i \ge 1,\ i' = i+1 \\ m\mu & \text{if } i' = i-1 \\ 0 & \text{in other cases,} \end{cases} \tag{2.80}$$

where

$$f := \sum_{k=m}^{m+n} \rho_0(k).$$

Then from relation (2.80) the ergodicity condition $\tilde{\nu}_o < 1$ of the model is found which corresponds exactly to the similar condition found for the macro-cell model. Upon meeting the ergodicity condition from (2.80) the stationary distribution of the merged model is found:

$$\pi(<i>) = f\tilde{\nu}_o^i \pi(<0>),\ i \geq 1, \tag{2.81}$$

where

$$\pi(<0>) = \frac{1-\tilde{\nu}_o}{1-\tilde{\nu}_o + f\tilde{\nu}_o}.$$

After some mathematical transformation the following relations for approximate calculation of the QoS metrics of the micro-cell model with an unlimited queue for o-calls are found:

$$P_h \approx \rho_0 (m+n)\, \pi(<0>) + E_B(\nu_h, n)\,(1-\pi(<0>)), \tag{2.82}$$

$$L_o \approx \frac{1-\pi(<0>)}{1-\tilde{\nu}_o}. \tag{2.83}$$

As in the previous case, for the micro-cell model QoS metrics can also be found when there is finite buffer $R, R < \infty$ for o-calls. Then at any structural and load parameter values there is a stationary regime in the system. Dropping well-known steps in solution of this problem, below are given the final formulae for approximate calculation of QoS metrics of the micro-cell model with a finite queue:

$$P_h(R) \approx \rho_0 (m+n)\, \pi_R(<0>) + E_B(\nu_h, n)\,(1-\pi_R(<0>)), \tag{2.84}$$

$$L_o \approx f\pi_R(<0>) \sum_{i=1}^{R} i\tilde{\nu}_o^i, \tag{2.85}$$

$$P_o(R) \approx \pi_R(<R>), \tag{2.86}$$

$$W_o \approx \frac{L_o(R)}{\lambda_o(1-P_o(R))}. \tag{2.87}$$

Here stationary distribution of the merged model is calculated from the following:

$$\pi_R(<i>) = f\tilde{\nu}_o^i \pi_R(<0>),\ i = 1, 2, \ldots, R, \tag{2.88}$$

where

$$\pi_R(<0>) = \frac{1-\tilde{\nu}_o}{1-(1-f)\,\tilde{\nu}_o - f\tilde{\nu}_o^{R+1}}.$$

2.2.3 Numerical Results

Let us first examine the results of numerical experiments for both schemes with isolated channel reservation. To keep it brief only models for the microcell are shown (in which $\lambda_o << \lambda_h$) in two series of experiments. In both series the input data is chosen as follows: $m+n=20$, $\lambda_o = 0.5$, $\lambda_h = 10$. Also in one series it is assumed that $\mu = 0.8$, in another – $\mu = 1$.

In Fig. 2.14 the dependency of function P_h on the number of reserved channels is shown in the HOPS scheme. It is seen from this figure that for given input data this function decreases in a particular range of value n, thereafter it increases. This fact has the following explanation. With the increase of n (i.e. decrease of m) loss intensity of h-calls from the Primary Group increases and consequently the intensity of these calls in the Secondary Group of channels increases. Thus, in Erlang's B-formula load ($\tilde{v}_h$) and number of channels (n) increases simultaneously and therefore, it is impossible to foresee the P_h function's behavior from the number of reserved channels in this scheme. In other words determination of the type of P_h function for concrete values of the model's parameters requires appropriate numerical experiments.

Dependency of L_o on the number of reserved channels is shown in Fig. 2.15. This function is an increasing function at constant intensity of input traffic, since increase of the number of reserved channels leads to decrease of the number of channels in Primary Group, i.e. average queue length of o-calls increases. The same type of dependency on the number of reserved channels is demonstrated by W_o.

Figures 2.16 and 2.17 demonstrate the dependency of QoS parameters of the model on the number of reserved channels in the HOSP scheme. It is seen from

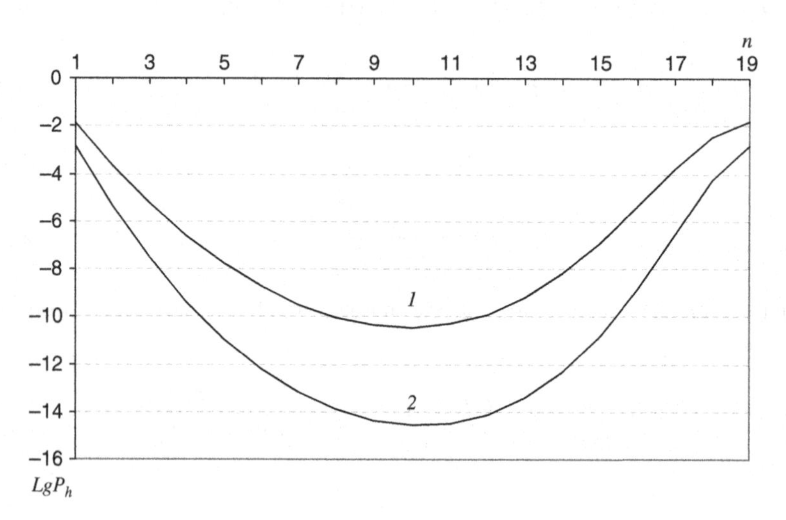

Fig. 2.14 P_h versus n for the HOPS model in the case where $m+n=20$, $\lambda_o=0.5$, $\lambda_h=10$; $1\mu=0.8$; $2–\mu=1$

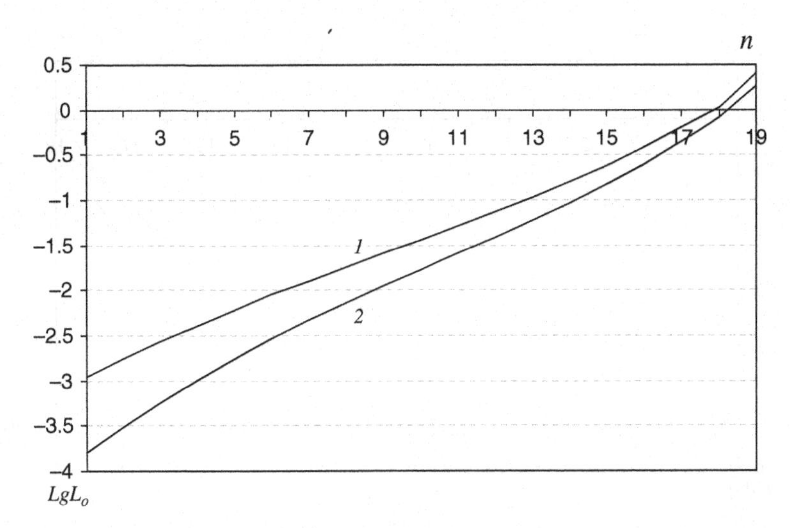

Fig. 2.15 L_o versus n for the HOPS model in the case where $m+n=20$, $\lambda_o=0.5$, $\lambda_h=10$; 1–$\mu=0.8$; 2–$\mu=1$

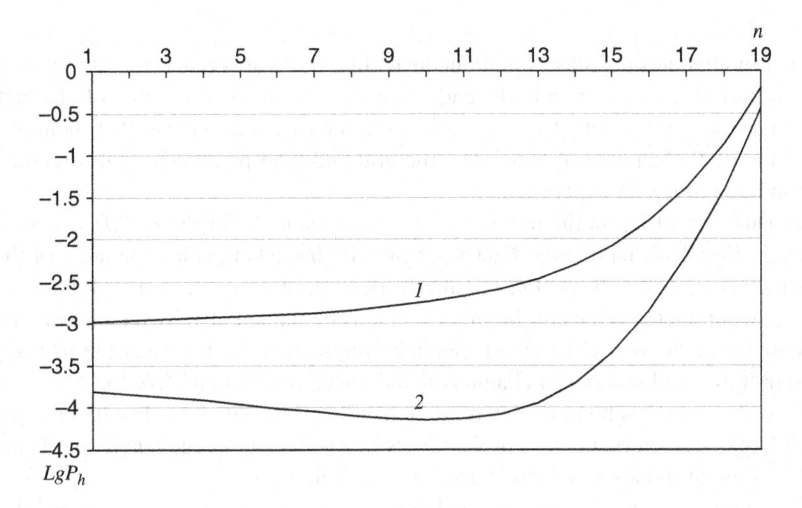

Fig. 2.16 P_h versus n for the HOSP model in the case where $m+n=20$, $\lambda_o=0.5$, $\lambda_h=10$; 1–$\mu=0.8$; 2–$\mu=1$

Fig. 2.16 that at $\mu=0.8$ the function P_h strictly increases in the whole range of the reserved channels value, whereas at $\mu=1$ it decreases within an interval [1, 10] and then increases again. Note that such behavior is specific to the given data and for other input data it will be different. As a matter of fact, in this scheme, the intensity of h-calls to the Primary Group of channels ($\hat{\lambda}_h$) decreases upon the increase of the number of reserved channels. At the same time, the number of channels in this group also decreases, i.e. in the queuing system $M/M/m/\infty$ both input traffic intensity and

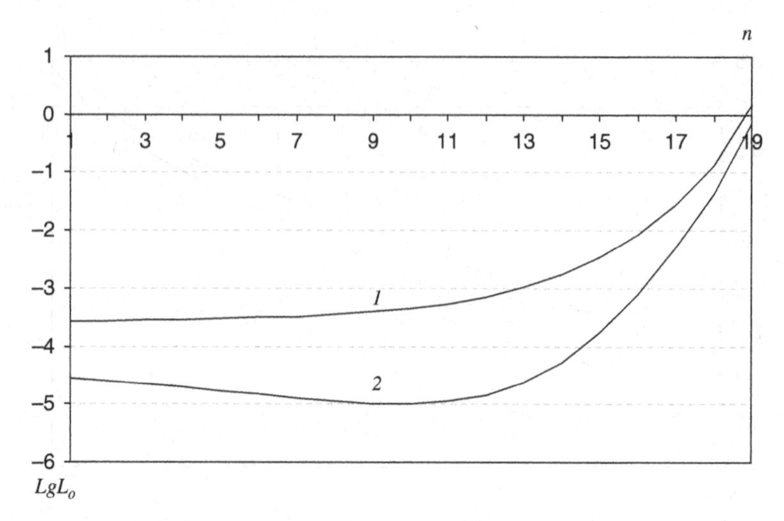

Fig. 2.17 L_{o} versus n for the HOSP model in the case where $m+n=20$, $\lambda_{\mathrm{o}}=0.5$, $\lambda_{\mathrm{h}}=10$; $1\text{–}\mu=0.8$; $2\text{–}\mu=1$

number of channels decrease simultaneously. In other words it is difficult to predict P_{h} behavior since it essentially depends on concrete values of number of channels as well as load parameters. Consequently, as in the case with the HOPS scheme the definition of P_{h} behavior at concrete structural and load parameter values requires appropriate numerical experiments.

Dependency of L_{o} on the number of reserved channels for the HOSP scheme is shown in Fig. 2.17. Unlike the HOPS scheme, in this scheme the monotony of this QoS parameter (as well as QoS parameter W_{o}) is not guaranteed at any load and structural parameter values of the model. This is explained also by the fact that the increase of the number of reserved channels simultaneously decreases the intensity of input traffic and number of channels in the queuing system $M/M/m/\infty$.

Note that in both schemes an increase of handling intensity has a beneficial effect on all QoS parameters, i.e. loss probabilities of h-calls and queue length and hence waiting time of o-calls (see Figs. 2.14, 2.15, 2.16, and 2.17).

Remarkably, from theoretical considerations we can deduce that loss probability of h-calls is less in the HOPS scheme than in the HOSP scheme, whereas queue length for o-calls is less in the HOSP scheme hence average waiting time is less than in HOPS. The first part of this clause is explained by the fact that the initial search of an empty channel in an "alien" group of channels increases the chances for h-calls to be accepted. The second part of the clause is explained by the fact that total load of incoming traffic in the Primary Group of channels in the HOPS scheme $(\lambda_{\mathrm{o}}+\lambda_{\mathrm{h}})$ is higher than that of the HOSP scheme $(\lambda_{\mathrm{o}}+\hat{\lambda})$. These are clearly demonstrated in Table 2.5.

From the table we can deduce that if the research aims at decreasing of loss probability of h-calls, then HOPS scheme has greater advantages compared to the

Table 2.5 Comparison of QoS metrics for the models HOPS and HOSP in the case where $m+n=20$, $\lambda_o=0.5$, $\lambda_h=10$, $\mu=0.8$

	P_h		L_o		W_o	
n	HOPS	HOSP	HOPS	HOSP	HOPS	HOSP
1	2.18945E-04	1.05607E-03	1.09198E-03	2.71982E-04	2.18397E-03	5.43963E-04
2	2.18945E-04	1.08889E-03	1.75272E-03	2.76861E-04	3.50543E-03	5.53721E-04
3	5.72726E-06	1.12561E-03	2.72991E-03	2.82620E-04	5.45982E-03	5.65239E-04
4	2.42487E-07	1.16770E-03	4.14018E-03	2.89725E-04	8.28036E-03	5.79451E-04
5	1.66361E-08	1.21737E-03	6.13771E-03	2.98889E-04	1.22754E-02	5.97777E-04
6	1.83783E-09	1.27779E-03	8.93135E-03	3.11207E-04	1.78627E-02	6.22414E-04
7	3.26134E-10	1.35421E-03	1.28126E-02	3.28418E-04	2.56252E-02	6.56836E-04
8	9.35392E-11	1.45476E-03	1.82008E-02	3.53361E-04	3.64017E-02	7.06723E-04
9	4.40913E-11	1.59290E-03	2.57188E-02	3.90880E-04	5.14376E-02	7.81759E-04
10	3.51240E-11	1.79172E-03	3.63195E-02	4.49701E-04	7.26389E-02	8.99401E-04
11	4.91568E-11	2.09295E-03	5.15097E-02	5.46673E-04	1.03019E-01	1.09335E-03
12	1.26751E-10	2.57717E-03	7.37537E-02	7.17194E-04	1.47507E-01	1.43439E-03
13	6.33316E-10	3.41292E-03	1.07233E-01	1.04361E-03	2.14466E-01	2.08721E-03
14	6.36998E-09	4.98949E-03	1.59309E-01	1.74254E-03	3.18619E-01	3.48508E-03
15	1.27549E-07	8.32246E-03	2.43389E-01	3.47529E-03	4.86779E-01	6.95058E-03
16	4.41134E-06	1.64976E-02	3.84356E-01	8.66709E-03	7.68711E-01	1.73342E-02
17	1.73348E-04	4.09669E-02	6.28121E-01	2.86297E-02	1.25624E+00	5.72594E-02
18	3.48161E-03	1.37682E-01	1.06891E+00	1.39320E-01	2.13781E+00	2.78640E-01
19	1.60324E-02	6.36581E-01	2.55009E+00	1.45224E+00	5.10018E+00	2.90447E+00

HOSP scheme, since in some individual cases this parameter is almost 10^8-times better. Likewise using the HOSP scheme allows decreasing average queue length for o-calls and hence the average waiting time of o-calls is almost 10^2-times shorter compared to the HOPS scheme.

This table also suggests a choice of appropriate scheme dependent on load and structural parameters of a model. The solution of such problems implies defining requirements for QoS parameters and thus finding a scheme that will meet these requirements. For instance, if for given selected input data one needs to meet the following requirement $P_h \leq 10^{-4}$, then the HOSP scheme will not allow doing this at any values of n, whereas in the HOPS scheme this requirement is met at $n=3,4,\ldots,16$.

It is important to note that the obtained results correspond exactly to those from [9], where the values were calculated using the fairly complex theory of a multi-dimensional generating function. Moreover, this work only suggests a solution for models with infinite queues of o-calls, whereas the suggested approach works with finite queues as well.

Numerical experiments were also conducted separately for macro-cell models (where $\lambda_o >> \lambda_h$) and for models with a symmetrical load of different types of calls (where $\lambda_o = \lambda_h$) with different schemes of channel assignment. For brevity these results are not given here. We will just note that the above-mentioned behavior of QoS parameters is demonstrated for any type of cell.

Now we will consider the results of numerical experiments for schemes with reassignment of channels. First we will consider the HRMA model. In numerical experiments for the model with an unlimited queue of o-calls initial data are chosen as follows: $N=40$, $\lambda_h=15$, $\mu=1$. As expected the function P_h (Fig. 2.18) decreases while both functions L_o (Fig. 2.19) and W_o (Fig. 2.20) increase versus

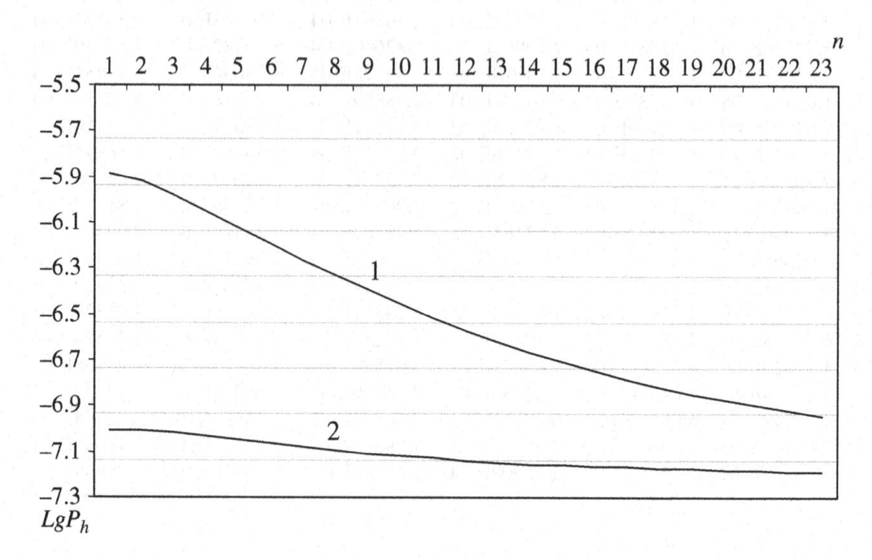

Fig. 2.18 P_h versus n for the HRMA model in the case where $N=40$, $\lambda_h=15$, $\mu=1$; $1\text{–}\lambda_o=4$, $2\text{–}\lambda_o=2$

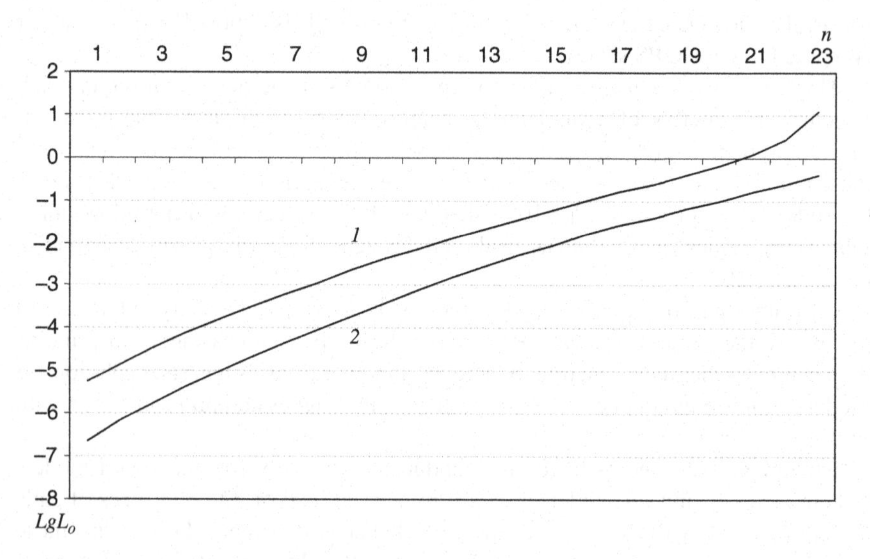

Fig. 2.19 L_o versus n for the HRMA model in the case where $N=40$, $\lambda_h=15$, $\mu=1$; $1\text{–}\lambda_o=4$, $2\text{–}\lambda_o=2$

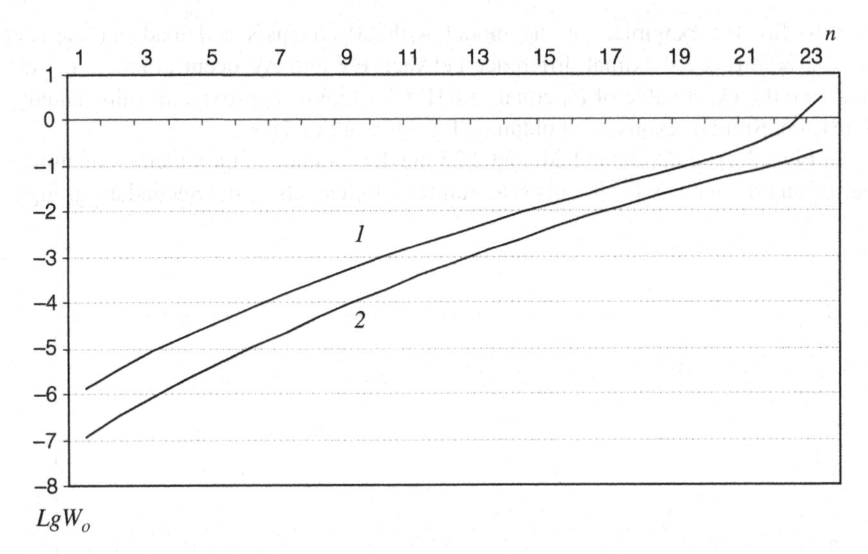

Fig. 2.20 W_o versus n for the HRMA model in the case where $N=40$, $\lambda_h=15$, $\mu=1$; $1\text{–}\lambda_o=4$, $2\text{–}\lambda_o=2$

the number of guard channels. Therewith all functions are increasing versus o-call load. Note that for the indicated initial data the ergodicity property of the model is violated at $n \geq 24$ therefore in graphs the values of parameter n are shown in interval [1, 23]. Numerical experiments were executed for the model with a limited queue of o-calls also and analogous results were found.

The accuracy of the proposed approximate formulae for the given model was also estimated. Exact values of QoS are considered to be those calculated by formulae which were proposed in [2]. Note that both the approximate and exact results for P_h are almost identical. Some results of the comparison are given in Table 2.6.

It is worth noting that sufficiently high accuracy exists even for the initial data not satisfying the known assumption concerning the ratios of traffic loads of o- and

Table 2.6 Comparison of exact and approximate values of QoS metrics for the model HRMA with patient o-calls in the case where $\lambda_o=1$, $\lambda_h=10$, $\mu=2.0$

		P_h	
$m+n$	n	EV	AV
20	3	8.43E-07	2.98E-07
20	5	7.02E-07	2.97E-07
20	7	5.88E-07	2.96E-07
30	3	1.92E-13	2.79E-14
30	5	1.59E-13	2.75E-14
30	7	1.32E-13	2.72E-14
30	20	4.07E-14	2.71E-14
40	7	1.07E-21	9.42E-23
40	20	3.17E-22	8.54E-23

h-calls. So, for example, for the model with 20 channels and load parameters $\lambda_o = \lambda_h = 7$, $\mu = 2$ maximal difference between EV and AV occur at $n = 1$, i.e. in this case the exact value of P_h equals 1.91E-05 while its approximate value equals 4.13E-06. Similar results were obtained for other initial data.

In Figs. 2.21, 2.22, 2.23, 2.24, and 2.25 the dependency of QoS metrics on number of guard channels for h-calls (i.e. number of channels in the Secondary group)

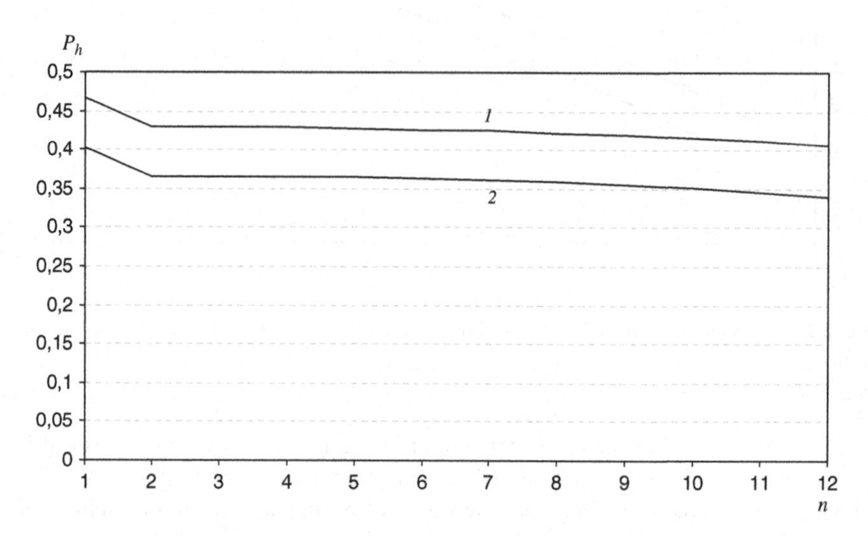

Fig. 2.21 P_h versus n for the HOPSWR model of the microcell in the case where $m+n = 15$, $\lambda_o = 5$, $\mu = 2$; 1–$\lambda_h = 40$, 2–$\lambda_h = 35$

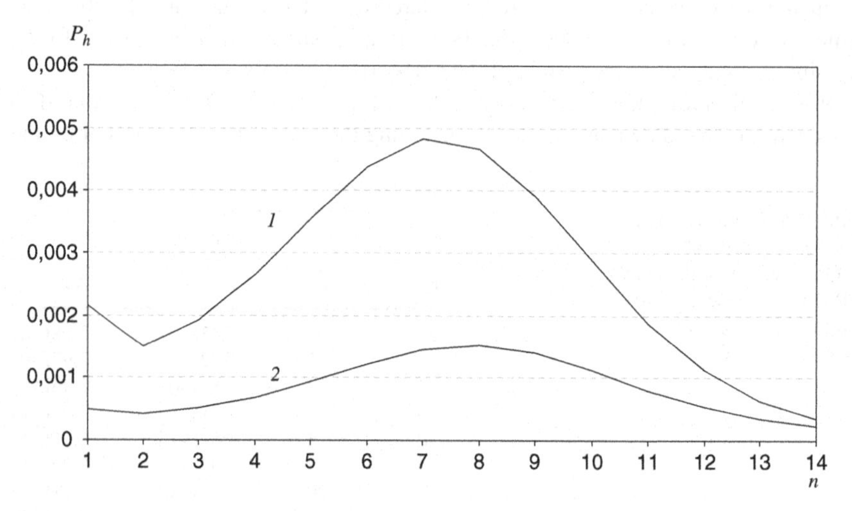

Fig. 2.22 P_h versus n for the HOPSWR model of the microcell in the case where $m+n = 15$, $\lambda_h = 20$, $\mu = 4$; 1–$\lambda_o = 5$, 2–$\lambda_o = 2$

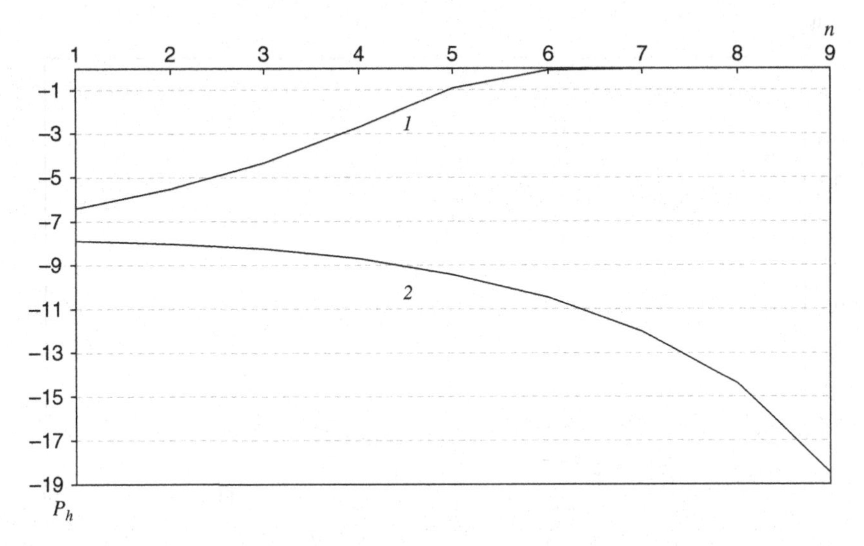

Fig. 2.23 P_h versus n for the HOPSWR model of the macrocell in the case where $m+n=10$, $\lambda_0=15$, $\mu=20$; 1–$\lambda_h=3$, 2–$\lambda_h=0.9$

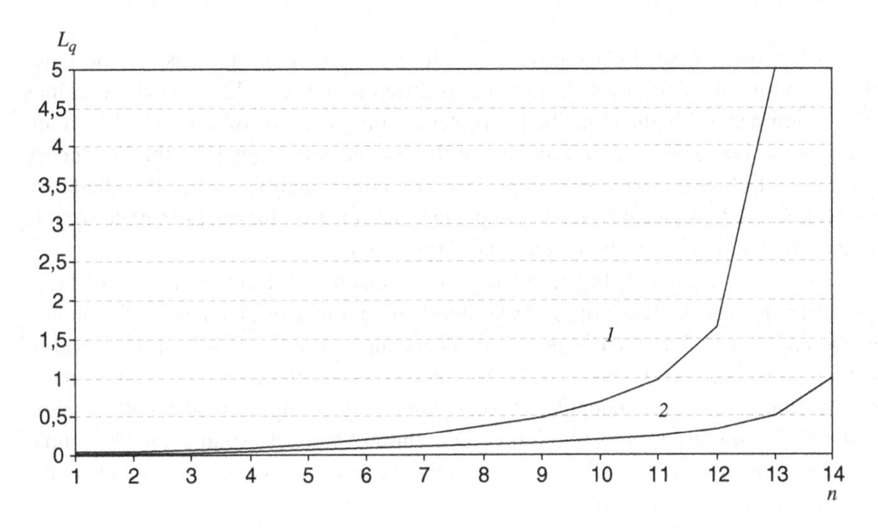

Fig. 2.24 L_q versus n for the HOPSWR model of the macrocell in the case where $m+n=15$, $\mu=2$; 1–$\lambda_o=5$, $\lambda_h=15$, 2–$\lambda_o=2$, $\lambda_h=20$

in the model HOPSWR are given. They completely meet theoretical expectations. So, the P_h function's shape in the microcell is given in Figs. 2.21 and 2.22. It can be seen from Fig. 2.21 that at fixed initial values of the model, this function systematically decreases. This is explained by the fact that at these values of cell parameters h-calls use channels of the Primary group poorly, i.e. they in fact use

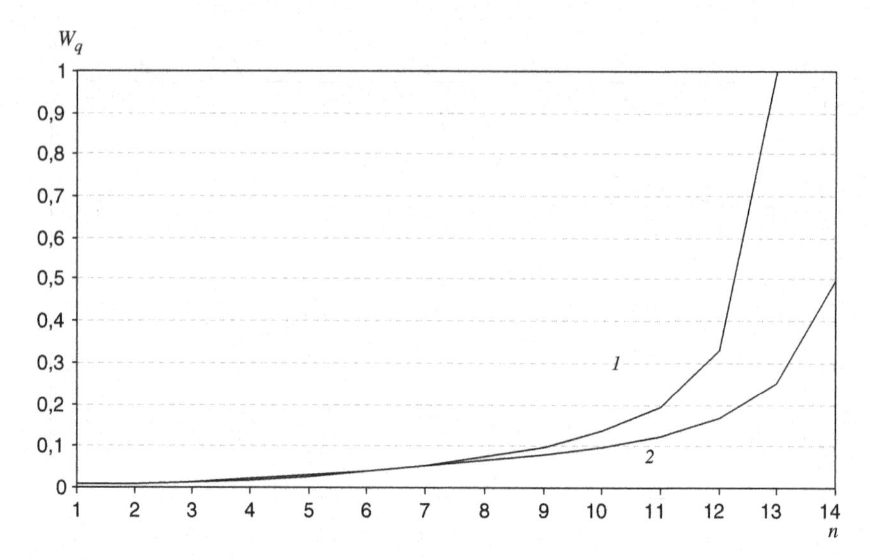

Fig. 2.25 W_q versus n for the HOPSWR model of the macrocell in the case where $m+n=15$, $\mu=2$; 1–$\lambda_o=5$, $\lambda_h=15$, 2–$\lambda_o=2$, $\lambda_h=20$

channels from the Secondary group, and, therefore, with the increase of the latter the given function decreases. The picture is different in Fig. 2.22. Here at low values of the number of channels in the Secondary group, h-calls use all available channels poorly, however with an increase of the number of channels in the Secondary group, total channel usage rate improves, and consequently, P_h has the shape we see in Fig. 2.22. Noticeably, as was expected, the function increases in a monotonic fashion depending on traffic loads of both types of calls.

Figure 2.23 depicts P_h function shape in the macrocell. Here both cases of function decrease and increase upon variation of the number of channels in Secondary group are shown. The load of new calls is constant. In other words, at low load h-calls mainly occupy channels in the Secondary group, therefore, with an increase of the number of such channels, P_h decreases. And at higher load h-calls occupy channels from both groups, but at the given initial values the total channel occupation rate of h-calls worsens, hence with an increase of the number of channels in the Secondary group P_h increases.

In both types of cells functions L_o and W_o increase with respect to the number of channels in the Secondary group regardless of traffic loads and total number of channels. Figures 2.24 and 2.25 depict these functions' shape for the macrocell. They have similar shapes for the microcell as well.

Analysis of numerical experiments reveals that the problem of optimal distribution of channels with the aim of compliance with QoS parameters of different types of calls is not a trivial one, consequently, its solution in each concrete case will require special investigation. This is caused by P_h behavior at various relations of traffic intensities (see Figs. 2.21, 2.22, and 2.23).

Table 2.7 Comparison of exact and approximate values of QoS metrics for the model HOPSWR with patient o-calls in the case where $m+n=15$, $\lambda_o=5$, $\lambda_h=15$, $\mu=2$

	P_h		L_q		W_q	
n	EV	AV	EV	AV	EV	AV
1	0.0509	0.0506	0.0401	0.0358	0.0098	0.0072
2	0.0445	0.0436	0.0179	0.0366	0.0088	0.0073
3	0.0498	0.048	0.0625	0.0609	0.0156	0.0122
4	0.0588	0.0534	0.0989	0.0945	0.0201	0.0189
5	0.0601	0.0583	0.1542	0.1395	0.0302	0.0279
6	0.0626	0.0612	0.2002	0.1978	0.0411	0.0396
7	0.0655	0.0611	0.2823	0.2724	0.0600	0.0545
8	0.0599	0.0576	0.3987	0.3678	0.0765	0.0736
9	0.0545	0.0511	0.5002	0.4943	0.0856	0.0989
10	0.0478	0.0428	0.6987	0.6757	0.1246	0.1351
11	0.0352	0.0341	0.9899	0.9792	0.2003	0.1958
12	0.0279	0.0259	1.6803	1.6576	0.2998	0.3315

Another aim of numerical experiments is to measure the accuracy of suggested formulae. Thus, the approximate results for macro- and microcells are almost completely identical to the results of [10] which are considered exact when $\mu_o=\mu_h$. Some comparisons for the microcell are given in Table 2.7. Noticeably, the accuracy of given formulae increases with the increase of intensity ratios of different types of calls. Similar results are also achieved for other parameters of the models studied.

It is important to note that the given approximate formulae have low accuracy at close values of load parameters of heterogeneous calls (i.e. when $\lambda_o \approx \lambda_h$ and $\mu_o \approx \mu_h$) and therefore cannot be used in QoS research for cells where load parameters of original and handover calls do not differ substantially.

2.3 Conclusion

In this chapter simple numerical procedures for calculation of QoS metrics in wireless networks are proposed, where well-known shared channel reservation schemes for prioritized h-calls and either limited or unlimited queues for homogenous calls are used. It is important to note that unlike the classical models of the above communication networks, here new and handover calls are assumed not to be identical in time of radio channel occupancy. The works [3, 5] applied the analytical models of a cell with an unlimited queue of h-calls under the assumption that the duration of the degradation interval has an exponential distribution. The analogous model with a limited queue of h-calls and infinite degradation interval was studied in [11]. The works [6, 7] suggested numerical algorithms for studying models with a limited length of queue for h-calls. Therewith consideration was given to models with patient [6] and impatient calls [7]. In [1] signal-flow graphs and Mason's formula were used to obtain the blocking probabilities of o- and h-calls and mean waiting times in the model with a limited queue of both kinds of calls and reneging/dropping of waiting calls. In all the above works o- and h-calls were assumed to be identical in terms of channel occupancy time. Computational procedures to calculate QoS metrics of investigated networks with an unlimited queue of patient or impatient h-calls were proposed in [4]. In the latter work it was assumed that o- and h-calls were nonidentical.

Models of investigated networks with unlimited queues of o-calls were studied in [2, 9]. For calculation of QoS metrics of the HRMA model, a matrix-geometric approach was used in [2], while generation of function method in conjunction with matrix spectral tools for the model HOPSWR was used in [9]. Note that these methods allow for calculation of QoS metrics in the case of an unlimited queue of o-calls only. Approximate methods to calculate QoS metrics for the model HOPSWR were proposed in [10]. It is important to note that unlike the methods that were proposed in [2, 9] the approximate method allows one to investigate models with limited queues of o-calls as well. The proposed method can also be used for research into two-dimensional models where more sophisticated channel reservation schemes are used as well as for models with a finite buffer for both kinds of impatient calls, for example see [1, 12]. Simple algorithms for computing the QoS metrics of HOPS and HOSP schemes for channel assignment were proposed in [8].

References

1. Chang CJ, Su TT, Chung YY (1994) Analysis of a cutoff priority cellular radio system with finite queuing and reneging/dropping. IEEE/ACM Trans Netw 2(2):166–175
2. Guerin R (1988) Queuing-blocking system with two arrival streams and guard channel. IEEE Trans Commun 36(2):153–163
3. Hong D, Rapoport SS (1986) Traffic model and performance analysis of cellular mobile radio telephone systems with prioritized and nonprioritized handoff procedures. IEEE Trans Veh Technol 35(3):77–92
4. Kim CS, Melikov AZ, Ponomarenko LA (2007) Two-dimensional models of wireless cellular communication network with infinite queues of h-calls. J Autom Inf Sci 39(12):25–41
5. Lin YB, Mohan S, Noerpel A (1994) Queuing priority channel assignment strategies for PCS handoff and initial access. IEEE Trans Veh Technol 43(3):704–712
6. Melikov AZ, Ponomarenko LA, Babayev AT (2005) Numerical method to study mobile cellular wireless network models with finite queue of h-calls. J Autom Inf Sci 37(6):1–11
7. Melikov AZ, Ponomarenko LA, Babayev AT (2006) Investigation of cellular network characteristics with limited queue of impatient h-calls. J Autom Inf Sci 38(8):17–28
8. Melikov AZ, Velibekov AM (2009) Computational procedures for analysis of two channel assignment schemes in wireless cellular networks. Autom Control Comput Sci 43(2):96–103
9. Pla V, Casares-Giner VA (2005) Spectral-based analysis of priority channel assignment schemes in mobile cellular communication systems. Int J Wirel Inf Netw 12(2):87–99
10. Ro IS, Melikov AZ, Ponomarenko LA, Kim CS (2008) Numerical approach to analysis of channel assignment schemes in mobile cellular networks. J Korea Manage Eng Soc 13(2): 29–39
11. Yoon CH, Un CK (1993) Performance of personal portable radio telephone systems with and without guard channels. IEEE J Sel Areas Commun 11(6):911–917
12. Zhuang W, Bensaou B, Chua KC (2000) Handoff priority scheme with preemptive, finite queueing and reneging in mobile multi-service networks. Telecommun Syst 15:37–51

Chapter 3
QoS Optimization Problems in Cellular Wireless Networks

The algorithms suggested in previous chapters for calculation of QoS (Quality of Service) metrics of integrated voice/data Cellular Wireless Networks (CWN) allow one to optimize their values in terms of defined QoS requirements. QoS requirements can be defined say by indicating upper bounds for the loss probabilities of original (o-calls) and handover calls (h-calls) of heterogeneous traffic as well as lower bounds for the number of busy channels. In this chapter some optimization problems for both cells without queues and with queues are considered. For the sake of simplicity we consider optimization problems for mono-service networks and we have attempted as much as possible to retain the notation from previous chapters.

3.1 QoS Optimization Problems in Cells Without Queues

The formulae obtained in Chap. 1 for approximate calculation of QoS metrics of cells without queues allow one to carry out their optimization with respect to selected performance criterion of handling of heterogeneous calls. Below some problems of this kind are considered.

3.1.1 Optimization of Models with Guard Channels for Handover Calls

First we consider optimization problems for cells with guard channels (shared reservation). Through use of the algorithm that was proposed in Chap. 1 for calculating $P_h(N, g)$ and $P_o(N, g)$ it is possible to solve the minimization problem for these quantities. Moreover, for fixed N, ν_h, and ν_o, it turns out that g (number of guard channels for handover calls) becomes the only controlled parameter.

Knowing the range of variation of both functions $P_h(N, g)$ and $P_o(N, g)$ proves to be extremely useful for solving the problem of minimizing these two quantities.

L. Ponomarenko et al., *Performance Analysis and Optimization of Multi-Traffic on Communication Networks*, DOI 10.1007/978-3-642-15458-4_3,

Since $P_o(N, g)$ is a monotonically increasing function of g while $P_h(N, g)$ is a monotonically decreasing function of g, the following un-improvable boundaries may be proposed for these quantities:

$$P_o(N,0) \le P_o(N,g) \le P_o(N,N-1), \tag{3.1}$$

$$P_h(N,N-1) \le P_h(N,g) \le P_h(N,0). \tag{3.2}$$

It can be seen from (3.1) and (3.2) that the fairest processing is achieved when $g = 0$, since in this case $P_o(N, 0) = P_h(N, 0)$. But least fair processing is achieved when $g = N - 1$, since in this case the difference $P_o(N, g) - P_h(N, g)$ reaches a maximum value, i.e.

$$\max_g \{P_o(N,g) - P_h(N,g)\} = P_o(N,N-1) - P_h(N,N-1). \tag{3.3}$$

The following shows one problem in which (extreme) values of the parameter g must be found (with fixed N) such that there is a specified level of dropping of h-calls subject to blocking of o-calls [1]. Suppose that for given N, ν_h, and ν_o an optimal value of g must be found so as to minimize $P_o(g)$ with a given constraint on $P_h(g)$, i.e. it is necessary to solve the following problem:

$$P_o(g) \underset{g}{\longrightarrow} \min, \tag{3.4}$$

$$\text{s.t. } P_h(g) \le \varepsilon_h, \tag{3.5}$$

where ε_h is a given value.

Herein to detect the optimization parameter of the considered problems in notations of QoS metrics we will explicitly indicate the regulated argument since the other parameters of the model are fixed.

Problem (3.4) and (3.5) is easily solved on the basis of (3.1) and (3.3) and the monotonic property of both P_h and P_o functions. In fact, in light of these remarks a minimal value $g^* \in [0, N-1]$ may be found using any one of the well-known methods of one-dimensional search (in particular, the dichotomy method) such that the constraint (3.5) is satisfied; it is precisely this value of g^* which is the solution of problem (3.4) and (3.5). Moreover, it is evident that if $\varepsilon_h < P_h(N-1)$, problem (3.4) and (3.5) will not have a solution.

A more interesting problem is finding such extreme values of g (with fixed N) that will provide the given level of QoS for different types of calls. At that, QoS level is often determined by the level of forced termination and blocking of new calls, respectively. One such possible setting comprises the following.

With fixed N, ν_h, and ν_o, it is required to find such ranges of value g, within which the given requirements for $P_h(g)$ and $P_o(g)$ are met, i.e. it is required find such $\underline{g}$, $\overline{g} \in [0, N-1]$, $\underline{g} \le \overline{g}$ so that

$$\overline{g} - \underline{g} \to \max, \tag{3.6}$$

$$\text{s.t. } P_o(g) \le \varepsilon_o, \tag{3.7}$$

$$P_h(g) \le \varepsilon_h, \ \forall g \in [\underline{g}, \overline{g}], \tag{3.8}$$

where ε_o and ε_h are given values.

The following algorithm can be proposed for a solution of (3.6), (3.7), and (3.8), taking into consideration (1.9) and (1.10) in Sect. 1.1.

Step 1. If $\varepsilon_h < P_h(N\text{–}1)$ and/or $\varepsilon_o < P_o(0)$ then this problem has no solution.
Step 2. In parallel, the following one-dimensional problems are solved:

$$g_o^* := \max\{g \,|P_o(g) \le \varepsilon_o\}, \tag{3.9}$$

$$g_h^* := \min\{g \,|P_h(g) \le \varepsilon_h\}. \tag{3.10}$$

Step 3. If $g_o^* < g_h^*$, then the problem (3.6), (3.7), and (3.8) has no solution; otherwise, the solution is $\underline{g} := g_h^*$, $\overline{g} := g_o^*$.

In step 2, for solution of problems (3.9) and (3.10), the well-known dichotomy method may be applied using the monotonic property of functions $P_o(g)$ and $P_h(g)$.

The results of a solution of the optimization problems (3.4), (3.5), (3.6), (3.7), and (3.8) in the case of $N = 100$, $\nu_o = 40$ Erl are shown in Tables 3.1 and 3.2, respectively. Herein the symbol Ø denotes that the corresponding problem has no solution.

It is evident from Table 3.1 that g^* decreases with increasing ε_h while g^* increases with increasing ν_h (for fixed values of the other parameters). These results were expected, since imposing stricter requirements on the level of losses of handover requests as well as increasing the load of these calls requires increasing the number of guard channels in order to service this type of call.

Table 3.1 Solution results for the problem (3.4) and (3.5) in the case where $N = 100$, $\nu_o = 40$ Erl

ν_h	40	40	40	40	40	50	50	50	50	50
ε_h	10^{-2}	10^{-3}	10^{-4}	10^{-5}	10^{-6}	10^{-2}	10^{-3}	10^{-4}	10^{-5}	10^{-6}
g^*	0	11	20	27	33	7	19	27	34	41

Table 3.2 Solution results for the problem (3.6), (3.7) and (3.8) in the case where $N = 100$, $\nu_o = 40$ Erl

ν_h	20	20	40	40	50	50	50	60	60	60
ε_o	10^{-2}	10^{-2}	$4{\cdot}10^{-1}$	$4{\cdot}10^{-1}$	10^{-2}	$5{\cdot}10^{-2}$	$5{\cdot}10^{-1}$	$5{\cdot}10^{-1}$	$5{\cdot}10^{-1}$	$5{\cdot}10^{-1}$
ε_h	10^{-7}	10^{-8}	10^{-3}	10^{-4}	10^{-3}	10^{-3}	10^{-3}	10^{-3}	10^{-2}	10^{-1}
$[\underline{g}, \overline{g}]$	[16,28]	[22,22]	[11,33]	[20,33]	Ø	Ø	[19,29]	Ø	[13,19]	[0,19]

From Table 3.2 it is evident that $\underline{g}$ decreases with increasing ε_h over the range of existence of the solution of problem (3.6), (3.7), and (3.8), while $\overline{g}$ also grows with increasing ε_o (with fixed values of the other parameters).

Now consider the following problem. It is required to find a maximal value of g such that given constraints to $P_o(g)$, $P_h(g)$ and in addition to $\tilde{N}(g)$ be satisfied. Formally this problem may be written as follows:

$$g \to \max, \tag{3.11}$$

$$\text{s.t. } P_o(g) \leq \varepsilon_o, \tag{3.12}$$

$$P_h(g) \leq \varepsilon_h, \tag{3.13}$$

$$\tilde{N}_o(g) \geq N_o \,, \tag{3.14}$$

where ε_o, ε_h, and N_o are given values.

The optimal solution of the problem (3.11), (3.12), (3.13), and (3.14), if it exists, is denoted g^*. To solve the given problem the following algorithm can be used.

Step 1. Find the maximal value of $g, g \in [0, N - 1]$ such that condition (3.14) is true. This value is denoted by g_o.
Step 2. If $P_h(g_o) > \varepsilon_h$ then the given problem has no solution.
Step 3. If $P_o(g_o) \leq \varepsilon_o$ then$g^* := g_o$.
Step 4. In $[0, g_o]$ find the maximal value of $\tilde{g}$ such that condition $P_o(\tilde{g}) \leq \varepsilon_o$ is true. If $P_h(\tilde{g}) \leq \varepsilon_h$ then $g^* := \tilde{g}$; otherwise the given problem has no solution.

The optimal solution of the problem (3.11), (3.12), (3.13), and (3.14) depends on values of six parameters, therefore, it is difficult to make any general conclusions concerning variation of g^* in response to the changes of initial data (see Table 3.3). However, there are some apparent outcomes, for instance, with increasing $\tilde{N}_{av}$ the optimal solution of the problem (if it exists) decreases.

The complexity of the developed algorithms for optimization problems is sufficiently low, since these algorithms require calculations of values of functions $P_o(g)$, $P_h(g)$, and $\tilde{N}(g)$ at most in $[\log_2 N]$ points (where $[x]$ – whole part of x) and comparison with given values ε_o, ε_h, and N_o. Thus, the complexity of these algorithms is logarithmic and therefore will not cause any difficulties in realization.

3.1.2 Optimization of Models with Individual Channels for Handover Calls

Now we consider the optimization problems for cells with individual channels (isolated reservation). In this model a parameter r, the number of private channels for

Table 3.3 Solution results for the problem (3.11), (3.12), (3.13), and (3.14)

ν_o	ν_h	ε_o	ε_h	N	N_o	g^*
25	25	10^{-2}	10^{-3}	80	48	13
30	35	10^{-2}	10^{-4}	100	60	16
35	30	10^{-3}	10^{-4}	100	60	8
40	40	10^{-3}	10^{-4}	120	70	10
40	40	10^{-2}	10^{-4}	120	70	19
45	50	10^{-3}	10^{-4}	150	70	23
45	50	10^{-2}	10^{-5}	150	80	23
60	50	10^{-3}	10^{-4}	150	80	6
60	50	10^{-3}	10^{-5}	120	80	Ø
40	40	10^{-2}	10^{-6}	120	70	Ø
40	40	10^{-2}	10^{-5}	120	75	19
65	70	10^{-3}	10^{-6}	120	75	Ø
70	45	10^{-3}	10^{-5}	150	100	Ø

h-calls, enters into the proposed scheme of prioritization of h-calls with a fixed number of service channels. It is sometimes possible to satisfy specified constraints on the characteristics of the system by varying the value of this parameter. An increase in r leads to a decrease in the loss probability of h-calls, but also increases the loss probability of o-calls as well as decreasing the utilization efficiency of the channels. The latter circumstance manifests itself particularly acutely in residential cells in which h-calls occur at a low rate.

In this light problems associated with the selection of extremal values of the parameter r that would not only satisfy constraints on the loss probability of different types of calls but would also maintain the utilization factor of the channels within a desirable range would be of interest.

Here we wish to consider one problem of this type. Suppose that the total number of channels N and the load parameters of different types of traffic ν_o and ν_h are specified quantities. Suppose also that constraints on the loss probability of different types of calls as well as on the utilization efficiency of the channels are also specified, i.e. these quantities must satisfy the following constraints:

$$P_o(r) \le \varepsilon_o\,, \tag{3.15}$$

$$P_h(r) \le \varepsilon_h\,, \tag{3.16}$$

$$\frac{1}{N}\tilde{N}(r) \ge \varepsilon_c\,, \tag{3.17}$$

where ε_o, ε_h, and ε_c are specified numbers.

The problem is to find the maximal ($\overline{r}$) and minimal ($\underline{r}$) values of the parameter r (if such exist) that satisfy the constraints (3.15), (3.16), and (3.17), i.e. it is necessary to solve the following problem:

$$\overline{r} - \underline{r} \rightarrow \max \tag{3.18}$$

under the constraints (3.15), (3.16), and (3.17), $\overline{r}, \underline{r} \in [0, N-1]$.

The problem may be divided into two separable problems.

Problem I. Find $\overline{r}$ under the constraints (3.15), (3.16), and (3.17), $\overline{r} \in [0, N-1]$.
Problem II. Find $\underline{r}$ under the same constraints.

If both problems simultaneously have solutions, the optimal solution of the initial problem (3.21), (3.22), (3.23), and (3.24) is constructed in an obvious way; otherwise (i.e. if at least one of these problems does not have a solution) the initial problem does not have a solution.

Now let us consider a method of solving the separable problems given above. Note that in solving these problems the monotonic properties of the investigated functions subject to the argument r for fixed N will be essentially used.

From condition (3.17) it is evident that if $\frac{1}{N}\tilde{N}(0) < \varepsilon_c$, the particular problem will not have a solution. In fact, since maximal utilization of the channels is achieved with $r = 0$, and if even with this technique (i.e. with the full-access technique) condition (3.17) does not hold, Problem I will not have a solution. Otherwise (i.e. if $\frac{1}{N}\tilde{N}(0) \geq \varepsilon_c$), an optimal solution of the problem (if it exists) is found in the closed interval $[0, r^*]$, where r^* is the maximal value of $r \in [0, N-1]$ such that condition (3.17) is satisfied. To find the value of r^* on the basis of the monotonic property of the function $\tilde{N}(r)$ we may apply the method of halving.

If $P_o(r^*) \leq \varepsilon_o$ and $P_h(r^*) \leq \varepsilon_h$, the solution of Problem I will be $\overline{r} := r^*$. If at the point $r = r^*$ condition (3.16) does not hold, then, independently of whether condition (3.15) is satisfied, Problem I will not have a solution. This is explained by the fact that since the function $P_h(r)$ decreases monotonically relative to the argument r and condition (3.16) is not satisfied at the point $r = r^*$, it may be satisfied only with $r > r^*$, whereas condition (3.17) is violated when $r > r^*$.

If $P_o(r^*) > \varepsilon_o$ and $P_h(r^*) \leq \varepsilon_h$, a maximal r^{**} may be found in the closed interval $[0, r^*]$ such that $P_o(r^{**}) \leq \varepsilon_o$. The method of dichotomy may be used to find r^{**} also in light of the fact that the function $P_o(r)$ is monotonic with respect to the argument r. Then, if $P_h(r^{**}) \leq \varepsilon_h$, the solution of Problem I will be $\overline{r} := r^{**}$; otherwise, the problem does not have a solution.

Summarizing, the following algorithm may be proposed for solving Problem I.

Step 1. If $\frac{1}{N}\tilde{N}(0) < \varepsilon_c$, Problem I does not have a solution.
Step 2. A maximal value r^* is found in the closed interval $[0, N-1]$ such that condition (3.17) is satisfied.
Step 3. If $P_h(r^*) > \varepsilon_h$, Problem I does not have a solution.

Step 4. If $P_o(r^*) \leq \varepsilon_o$ and $P_h(r^*) \leq \varepsilon_h$, the solution of Problem I will be $\overline{r} := r^*$.

Step 5. In the closed interval $[0, r^*]$ a maximal r^{**} is found such that $P_o(r^{**}) \leq \varepsilon_o$. If $P_h(r^{**}) \leq \varepsilon_h$, the solution of Problem I will be $\overline{r} := r^{**}$; otherwise, the problem does not have a solution.

Using this technique of solving Problem I, the following algorithm for solving Problem II may be proposed.

Step 1. Same as Step 1 of the solution algorithm for Problem I.
Step 2. Same as Step 2 of the solution algorithm for Problem I.
Step 3. Same as Step 3 of the solution algorithm for Problem I.
Step 4. If $P_o(r^*) \leq \varepsilon_o$ and $P_h(r^*) \leq \varepsilon_h$,a minimal r^{**} is found in the closed interval $[0, r^*]$ such that $P_h(r^{**}) \leq \varepsilon_h$. Then the solution of Problem II will be $\underline{r} := r^{**}$.
Step 5. In the closed interval $[0, r^*]$ a minimal r^{**} is found such that $P_h(r^{**}) \leq \varepsilon_h$. If $P_o(r^{**}) \leq \varepsilon_o$, the solution of Problem II will be $\underline{r} := r^{**}$; otherwise, the problem does not have a solution.

Obviously, problem (3.15), (3.16), (3.17), and (3.18) has a solution only if both Problem I and Problem II simultaneously have solutions, i.e., if at least one of these problems does not have a solution, the initial problem will also not have a solution.

Results of the optimization problem for (3.15), (3.16), (3.17), and (3.18) are shown in Table 3.4.

Table 3.4 Solution results for the problem (3.15), (3.16), (3.17), and (3.18) in the case where $N = 40$

ν_o	ν_h	ε_o	ε_h	ε_c	$[\underline{r}, \overline{r}]$
10	10	10^{-2}	10^{-2}	10^{-1}	[10,22]
50	35	10^{-2}	10^{-2}	10^{-1}	Ø
20	15	10^{-3}	10^{-1}	10^{-2}	Ø
15	5	10^{-2}	10^{-5}	10^{-1}	[12,16]
15	8	10^{-1}	10^{-4}	10^{-3}	[7,22]
20	8	10^{-1}	10^{-4}	10^{-3}	Ø
15	12	10^{-1}	10^{-4}	10^{-3}	Ø
20	5	10^{-1}	10^{-3}	10^{-2}	[8,17]
20	5	10^{-2}	10^{-3}	10^{-2}	[8,10]
20	5	10^{-2}	10^{-2}	10^{-2}	[0,10]
20	5	10^{-2}	10^{-3}	10^{-3}	[8,10]
5	10	10^{-2}	10^{-3}	10^{-3}	[0,29]
5	15	10^{-3}	10^{-4}	10^{-3}	[0,26]
5	15	10^{-3}	10^{-5}	10^{-3}	Ø

Solution of the given problem depends on the values of five parameters (under fixed values of number of channels) therefore as in the problem (3.11), (3.12), (3.13), and (3.14), in this case it is also difficult to make general conclusions concerning the location of optimal interval $[\underline{r}, \overline{r}]$ in response to the changes of initial data. However, there are some apparent outcomes, for instance, with increasing $\varepsilon_o(\varepsilon_h)$ the optimal value of $\overline{r}$ $(\underline{r})$ increases (decreases).

3.2 QoS Optimization Problems in Cells with Queues

The formulae obtained in Chap. 2 for approximate calculation of QoS metrics of cells with queues allow one to carry out their optimization with respect to selected performance criterion of handling the heterogeneous calls. Below some problems of this kind are considered.

3.2.1 QoS Optimization Problems in Cells with a Limited Queue of h-Calls

First we consider the QoS optimization problems for cells with a limited queue of patient h-calls. In this model it is admitted that h-calls in the queue were absolutely tolerant, i.e. there are no constraints on the time of their waiting in a queue. Here we consider the technique which allows us to take into account not only these constraints but constraints on loss probability of heterogeneous calls. The technique proposed is based on buffer size variation for waiting h-calls.

As already noted in Chap. 2, as the buffer size grows the average waiting time of h-calls in the system goes up as well, i.e. by choosing the appropriate value of parameter B one can make the values of the above characteristic (i.e. W_h) to be within desired limits. On the other hand, a decrease of parameter value B leads to an increase of the loss probability of h-calls (and decrease of the loss probability of o-calls). Thus, there appears the following optimization problem of the given model: for fixed loads of heterogeneous calls and for known values of total and reserved number of channels one needs to find the maximal value of buffer size $(\overline{B}^*)$ such that the prescribed constraints on the waiting time of h-calls in a queue as well as those on loss probability of heterogeneous calls are satisfied. Formally this problem is written as follows:

$$B \to \max \tag{3.19}$$

$$\text{s.t. } W_h(B) \le \delta, \tag{3.20}$$

$$P_h(B) \le \varepsilon_h, \tag{3.21}$$

$$P_o(B) \le \varepsilon_o, \tag{3.22}$$

where δ, ε_h and ε_o are the given values.

We move on to describing one of the possible algorithms for solution of problem (3.19), (3.20), (3.21), and (3.22). The first step determines the maximal value B so as to satisfy the constraint (3.20). Denote this value by $\overline{B}$. Note that to find $\overline{B}$ one can employ the following technique. The function W_{h} is approximated from below by the quantity $\underline{W}_{\text{h}}$ representing the average waiting time in the classical queuing system $M|M|N|B$ with the load ν_{h} Erl to find the least integer solution (with respect to B) of inequality $\underline{W}_{\text{h}} \le \delta$ denoted by B_0. Further in the interval $[1,\, B_0]$ we find the required $\overline{B}$. Therefore, the monotonic property of function $W_{\text{h}}(B)$ allows us to employ the dichotomy method.

Then if $P_{\text{h}}\left(\overline{B}\right) > \varepsilon_{\text{h}}$, the problem (3.19), (3.20), (3.21), and (3.22) has no solution since in this case to satisfy the condition (3.21) it is necessary to increase the values B over $\overline{B}$, which violates the condition (3.20). Otherwise (i.e. for $P_{\text{h}}\left(\overline{B}\right) \le \varepsilon_{\text{h}}$), if $P_{\text{o}}\left(\overline{B}\right) \le \varepsilon_{\text{o}}$, then $\overline{B}^* := \overline{B}$ is the solution of the problem (3.19), (3.20), (3.21), and (3.22); if $P_o\left(\overline{B}\right) > \varepsilon_o$, then from the interval $[1, \overline{B}]$ we determine the maximal $\tilde{B}$ so that $P_{\text{o}}\left(\tilde{B}\right) \le \varepsilon_{\text{o}}$. Then if $P_{\text{h}}\left(\tilde{B}\right) \le \varepsilon_{\text{h}}$, $\overline{B}^* := \overline{B}$ is the optimal solution of the problem (3.19), (3.20), (3.21), and (3.22); if $P_{\text{h}}\left(\tilde{B}\right) > \varepsilon_{\text{h}}$ then the problem (3.19), (3.20), (3.21), and (3.22) has no solution.

In an analogous way one can solve the problem of finding the minimal value B (i.e. $\underline{B}^*$) for which the constraints (3.20), (3.21), (3.22) are satisfied. Thus, by combining the solution of these two optimization problems one can point to the interval where the values of B are changed and within which the given constraints on QoS metrics of heterogeneous calls are satisfied.

Some results of solving the indicated optimization problem (3.19), (3.20), (3.21), and (3.22) are illustrated in Table 3.5 where Ø denotes that with the given set of parameters the problem has no solution.

Analysis of the results of the optimization problems considered allows us to draw some general conclusions:

- with any type of traffic load increase (concurrently or taken separately) the value $\overline{B}^*$ (if it exists) goes down;
- with an increasing radio channel rate the value $\overline{B}^*$ also goes up;
- with a weakening of the requirements concerning loss probabilities of different types of calls (i.e. with values ε_{o} and ε_{h} increasing) the interval length $[\underline{B}^*, \overline{B}^*]$ increases;
- with a weakening of the requirements concerning waiting time of h-calls (i.e., with δ increasing) the interval length $[\underline{B}^*, \overline{B}^*]$ decreases.

Now we consider the QoS optimization problems in cells with a limited queue of impatient h-calls. As was shown in the numerical experiments in Sect. 2.1.3 an increasing number of reserved channels at fixed values of other parameters of the network also increases the probability of blocking o-calls and simultaneously decreases the probability of losing h-calls. Hence, by increasing the number of reserved channels one can attain the desirable level of losses for major h-calls. However, this leads to inefficient use of the entire pool of channels, since the

Table 3.5 Solution results for the problem (3.19), (3.20), (3.21), and (3.22)

λ_o	λ_h	μ	ε_o	ε_h	δ	$\left[\underline{B}^*, \overline{B}^*\right]$
30	50	2	0.9	2E-01	0.6	[2,4]
30	50	2	0.9	E-01	2	Ø
10	40	2	0.7	E-02	2	[13,17]
10	40	3	0.2	E-04	2	[12,100]
60	25	3	0.6	E-03	0.35	[21,100]
60	25	3	0.5	E-02	0.34	[4,14]
60	25	2	0.8	E-02	0.55	[11,19]
60	25	1	0.8	E-02	0.55	Ø
50	15	1	0.9	E-02	1.3	[22,30]
50	15	1	0.9	E-01	1.4	[3,54]
50	15	2	0.55	E-02	0.5	Ø
50	15	1	0.9	E-03	1.5	[84,90]
65	40	3	0.95	E-04	2	[51,100]
65	40	3	0.95	E-03	3	[23,100]
65	40	2	0.9	E-01	0.7	[5,33]
65	40	2	0.9	5E-02	0.7	[11,33]
40	40	2	0.85	E-02	0.75	[26,93]
50	40	3	0.85	E-02	0.35	[8,16]
50	20	3	0.85	E-02	0.35	[2,100]
50	20	3	0.85	E-03	0.35	[7,100]

increasing number of reserved channels decreases the coefficient of utilization of radio channels. The latter circumstance is especially drastically revealed in residential cells where the intensity of h-calls is low.

Thus, there appears the following problem of improving QoS metrics of heterogeneous calls: under the known loading and structural parameters of the cell one needs to find an interval of variation of values of reserved channel number within which are satisfied the prescribed limitations on probabilities of losses of heterogeneous calls as well as the average number of busy channels.

This problem solution is obtained by solving the following separable optimization problems:

$$\overline{g} := \arg \max_{g \in [1,N]} \left\{ P_h(g) \le \varepsilon_h, P_o(g) \le \varepsilon_o, \tilde{N}(g) \ge N_o \right\}, \tag{3.23}$$

$$\underline{g} := \arg \min_{g \in [1,N]} \left\{ P_h(g) \le \varepsilon_h, P_o(g) \le \varepsilon_o, \tilde{N}(g) \ge N_o \right\}, \tag{3.24}$$

where ε_h, ε_o and N_o are the prescribed quantities.

It is evident that the problems (3.23) and (3.24) simultaneously either have solutions or do not. Considering the monotony of functions $P_h(g)$, $P_o(g)$, and $\tilde{N}(g)$ involved in problems (3.23) and (3.24) one can suggest the following algorithm of their solution. To be more specific we consider the algorithm of solving the problem (3.23).

If $\tilde{N}(1) \le N_o$, then the given problem (3.23) has no solution; otherwise we find the solution of the following problem:

$$g_1 := \arg \max_{g \in [1,N]} \left\{ \tilde{N}(g) \ge N_o \right\}. \tag{3.25}$$

If $P_h(g_1) > \varepsilon_h$, then the given problem (3.23) has no solution; otherwise if $P_o(g_1) \le \varepsilon_o$, then the solution to problem (3.23) is $\overline{g} := g_1$.

If $P_o(g_1) > \varepsilon_o$, the following problem

$$g_2 := \arg \max_{g \in [1,g_1]} \left\{ P_o(g) \le \varepsilon_o \right\}, \tag{3.26}$$

is being solved.

Then if $P_h(g_2) \le \varepsilon_h$, then $\overline{g} := g_2$ is the solution of the problem (3.23); otherwise problem (3.23) has no solution. Note, that for solving problems (3.25) and (3.26) one can employ the dichotomy method.

In an analogous way problem (3.24) is solved. By combining solutions of these two optimization problems one can indicate the ends of the variation interval of values g within which are satisfied the prescribed limitations on QoS metrics of heterogeneous calls.

Some results of solving the problems (3.23) and (3.24) are depicted in Table 3.6.

Further we draw some conclusions based on analyzing the results of the problems (3.23), (3.24):

- with reduced requirements on loss probability (i.e., with ε_o and ε_h growing), g also increases and g decreases;
- with reduced requirements on the average number of busy channels (i.e., with N_o growing) the size of the optimal interval is reduced and finally the problem has no solution;
- with decreased traffic of any type (i.e. λ_o and λ_h) the size of the optimal interval grows.

Table 3.6 Solution results for the problem (3.23) and (3.24) for $N = 40$, $B = 10$, $\gamma = 20$, $\mu = 5$

λ_o	10	10	10	15	20	20	5	5	5	6	4
λ_o	8	8	8	8	8	8	10	10	10	10	6
ε_h	E-02	E-04	E-06	E-03	E-02	E-02	1.5 E-02	E-01	E-01	E-07	E-04
ε_o	E-02	E-03	E-06	1.5 E-04	E-03	E-03	3 E-03	3 E-03	3 E-03	3 E-03	3 E-02
N_o	2	3	3	4	4	6	2	2	3	2	2
$[\underline{g}, \overline{g}]$	[1,30]	[1,27]	[1,24]	[27,27]	[27,28]	Ø	[32,32]	[32,34]	[32,33]	[34,34]	[1,18]

3.2.2 *QoS Optimization Problems in Cells with an Unlimited Queue of h-Calls*

The numerical experiments concerning calculation of the models with an unlimited queue of h-calls showed that on selecting the appropriate values of the number of guard channels one could attain the desired level of service. In other words, if the limitations on QoS metrics are given, one can find the values of the number of guard channels at which such limitations are satisfied. Further, we consider two problems for the model with an unlimited queue of h-calls.

In the maximization problem of utilization of network channels under the given limitations on loss probability of o-calls and waiting time of h-calls one should find a value of guard channel number such that the given limitations on the probability of o-call loss and the waiting time of h-calls would be fulfilled with the average number of busy channels being maximally feasible. Mathematically we present this problem as follows:

$$\tilde{N}(g) \to \max \tag{3.27}$$

$$\text{s.t. } P_o(g) \le \varepsilon_o, \tag{3.28}$$

$$W_h(g) \le \delta, \tag{3.29}$$

where ε_o and δ are the given quantities.

Considering the monotonic properties of functions $\tilde{N}(g)$, $P_o(g)$, and $W_h(g)$ for solving the problem (3.27), (3.28), and (3.29) we can propose the following algorithm.

Step 1. If $P_o(1) > \varepsilon_o$ or $W_h(N{-}1) > \delta$, then the problem has no solution.
Step 2. In parallel the following problems are being solved:

$$g_p := \arg\max\{P_o(g) \le \varepsilon_o | g = 1, 2, \ldots, N-1\}$$
$$g_w := \arg\min\{W_h(g) \le \delta | g = 1, 2, \ldots, N-1\}$$

Step 3. If $g_p < g_w$, then the problem has no solution. Otherwise, g_w is the solution of the problem.

Note that on Step 2 the above problems can be solved by the known dichotomy method.

Likewise for the problem (3.23) and (3.24) of the model of a cell with a limited queue of impatient h-calls here we consider the following problem. Let the limitations on all QoS metrics be given and one should find a variation interval (of maximal length) of values of guard channel number that these limitations would satisfy. Mathematically we present this problem as follows:

$$\overline{g} - \underline{g} \to \max, \tag{3.30}$$

$$\text{s.t. } P_o(g) \le \varepsilon_o, \tag{3.31}$$

$$W_h(g) \le \delta, \tag{3.32}$$

$$\tilde{N}(g) \ge N_o. \tag{3.33}$$

Taking into account the monotonic properties of functions $P_o(g)$, $W_h(g)$, and $N_{av}(g)$ for solving the problem (3.30), (3.32), and (3.33) we can suggest the following algorithm.

Step 1. If $P_o(1) > \varepsilon_o$ or $W_h(N{-}1) > \delta$ or $N_{av}(1) < N_o$ the problem has no solution.
Step 2. In parallel the following problems are being solved:

$$\begin{aligned} g_p &:= \arg\max\{P_o(g) \le \varepsilon_o | g = 1,2,\ldots,N-1\} \\ g_w &:= \arg\min\{W_h(g) \le \delta | g = 1,2,\ldots,N-1\} \\ g_{av} &:= \arg\max\left\{\tilde{N}(g) \ge N_o | g = 1,2,\ldots,N-1\right\} \end{aligned}$$

Step 3. If $min\{g_p, g_{av}\} < g_w$, the problem has no solution. Otherwise we pass to the next step.
Step 4. If $g_{av} < g_p$ the optimal problem solution will be $\underline{g} := g_w$, $\overline{g} := g_{av}$; otherwise – $\underline{g} := g_w$, $\overline{g} := g_p$.

Some results from solving the problems (3.27), (3.28), (3.29), (3.30), (3.31), (3.32), and (3.33) are presented in Tables 3.7 and 3.8, respectively.

Table 3.7 Solution results for the problem (3.27), (3.28), and (3.29) for $N = 20$, μ_o=1

λ_o	λ_h	μ_h	ε_o	δ	g^*
10	15	16	E-01	E-04	2
10	15	16	E-02	E-04	Ø
8	15	16	E-01	E-05	2
8	15	16	E-01	E-04	2
11	15	16	2.0E-01	E-06	6
11	15	16	2.0E-01	E-07	7
11	15	16	2.0E-01	E-08	Ø
11	15	20	E-01	E-06	5
11	15	20	E-01	E-05	4
11	15	20	E-01	E-04	2
11	15	20	E-01	E-03	1
11	15	20	E-02	E-03	Ø
11	10	15	E-01	E-05	4
11	10	15	E-02	E-05	Ø
11	10	15	E-01	E-06	5

Table 3.8 Solution results for the problem (3.30), (3.31), (3.32), and (3.33) for $N=20$, $\mu_o=1$

λ_o	λ_h	μ_h	ε_o	N_o	δ	$[\underline{g}, \overline{g}]$
10	15	16	E-01	10	E-04	[2,6]
10	15	16	E-02	10	E-04	Ø
8	15	16	E-01	5	E-05	[2,8]
8	15	16	E-01	5	E-04	[2,8]
11	15	16	2.0E-01	5	E-06	[6,7]
11	15	16	2.0E-01	5	E-07	[7,7]
11	15	16	2.0E-01	5	E-08	Ø
11	15	20	E-01	5	E-06	[5,5]
11	15	20	E-01	5	E-05	[4,5]
11	15	20	E-01	10	E-04	[2,5]
11	15	20	E-01	10	E-03	[1,5]
11	15	20	E-02	10	E-03	Ø
11	10	15	E-01	10	E-05	[4,5]
11	10	15	E-02	10	E-05	Ø
11	10	15	E-01	10	E-06	[5,5]

3.3 Conclusion

In this chapter simple numerical procedures for optimization of QoS metrics in wireless networks are proposed. For this both models with and without queues are considered. For the models without queues the problems of finding the optimal values of shared and isolated channel reservation strategy parameters are solved. Optimization of QoS metrics for the cell with a shared channel reservation scheme for priority h-calls and their limited or unlimited queue are also considered. Consideration has been given to models with patient and impatient h-calls. For the model with a limited queue of patient h-calls the problem of selecting the optimal size of the buffer is solved while for other models optimal values of the number of guard channels providing the given QoS level of heterogeneous calls are determined.

Note that optimization problems are solved at fixed values of load parameters of the network. However, as is known, the traffic intensities are the variable quantities and so of urgency becomes the problem of determining the intervals of invariance of the solution obtained, i.e. one should establish the intervals over which even with changing network load parameter values the solution obtained remains the same. It is important to note that when using the suggested algorithms there exists the possibility to conduct a numerical analysis of optimal solution sensitivity with respect to changing values of network load parameters.

Note, that the problems on optimization of QoS metrics of cells are insufficiently studied. This chapter is based on the results of the papers [2–4, 6]. Some similar problems were considered in [1, 5, 7].

References

1. Haring G, Marie R, Puigjaner R, Trivedi K (2001) Loss formulas and their application to optimization for cellular networks. IEEE Trans Veh Technol 50:664–673
2. Kim CS, Melikov AZ, Ponomarenko LA (2007) Two-dimensional models of wireless cellular communication networks with infinite queues of h-calls. J Autom Inf Sci 39(12):25–41
3. Melikov AZ, Babayev AT (2006) Refined approximations for performance analysis and optimization of queuing model with guard channels for handovers in cellular networks. Comput Commun 29(9):1386–1392
4. Melikov AZ, Ponomarenko LA, Babayev AT (2005) Numerical methods for investigation of cellular communication networks with finite queues of handover-calls. J Autom Inf Sci 37(6):1–11
5. Oh SH, Tcha DW (1992) Prioritized channel assignment in a cellular radio network. IEEE Trans Commun 40(7):1259–1269
6. Ponomarenko LA, Melikov AZ, Babayev AT (2006) Investigation of cellular network characteristics with limited queue of impatient h-calls. J Autom Inf Sci 38(8):17–28
7. Ramjee R, Towsley D, Nagarajan R (1997) On optimal call admission controls in cellular networks. Wirel Netw 3:29–41

Part II
Multi-Dimensional Models of Multi-Service Networks

Introduction

Modern telecommunication systems are multi-service systems, since they are handling a wide variety of messages – speech, video, data, fax, e-mail etc. The high complexity of these systems requires the development of adequate mathematical models with the aim of reliable evaluation and optimization of its characteristics. The main mathematical tools, allowing the development of analytical models that are adequate for multi-service telecommunication systems, are queuing systems and network theory. However, it is apparent that classic single-flow queuing systems can not serve as adequate mathematical models of real systems, since such single-flow queuing systems imply that all calls have identical parameters (for example, load parameters, priorities, handling mechanisms, etc.).

Indeed, in modern multi-service telecommunication networks calls are considerably different in terms of almost all parameters. Therefore, these systems can be described relatively reliably by multi-flow queuing systems. Classic multi-flow queuing systems are based on many prerequisites, one of which is essential: "one call – one channel". In other words, it is assumed, that a call of any type during the whole period of handling occupies only one channel (resource) of the system.

However, in modern multi-service telecommunication networks this precondition is not true. Since, for example, in such networks video information requires a wider bandwidth in the digital channel than for example audio or data. In the literature on teletraffic theory, such calls requiring many channels are called wide-band, whereas calls requiring lesser numbers of channels are called narrow-band. In the literature on queuing theory, multi-flow systems in which different types of calls require random numbers of channels are called Multi-Rate Queues (MRQ).

In MRQ it is required that two types of call are distinguished: elastic and inelastic. Inelastic calls are those that have fixed width. This means that upon handling of the inelastic call, all channels serving this call are occupied by this call at the same moment and any channel from the set of channels serving this call, if finished earlier, is blocked until all other channels from this set are also free. In other words, an inelastic call can only be served when there are enough free channels to serve it.

As opposed to inelastic calls, elastic calls do not have fixed width, i.e. for elastic calls only width ranges are predefined. This also means that the beginning and/or end of serving an elastic call can be different on different channels. In other words, upon handling of an elastic call the handling start time for different channels may be different and therefore every channel has its own individual handling jobs for this call and after finishing this job the channel is free for other calls.

Such multi-service telecommunication systems, in which both elastic and inelastic calls are handled, really exist. Such systems are called mixed.

Since in MRQ with inelastic calls the call will not be accepted until there are the required number of free channels, it is expected that wide-band calls will be blocked more often than narrow-band ones. Therefore, in such systems in order to maintain QoS metrics of heterogeneous calls at a required level an appropriate Call Admission Control (CAC) strategy must be defined.

Any access strategy in MRQ with pure loss CAC defines rules for call acceptance. These rules are based on various alternatives: whether to accept or reject the newly arrived call; whether or not to interrupt any existing call(s) in order to handle the newly arrived call, etc.

The simplest CAC is the fixed access strategy. When this strategy is used the alternatives to choose from are based on pre-defined rules that do not take into account the current state of the system, it being known that the state of the system can be described in various ways. It is worth noting, that practically speaking the use of a fixed CAC is preferable due to less requirements on software upon realization.

However, a real-life MRQ operates under considerable uncertainty of parameters of incoming traffic, which makes the use of a fixed CAC in many cases ineffective. These circumstances increase the necessity to research MRQ with controlled CAC. With the latter CAC alternatives to accepting or blocking incoming calls are based on the current state of the system.

Other classifications of MRQ with pure loss are also possible. However, taking into account the problems researched in this book we will confine ourselves to the above-mentioned classification principles. This part of the book covers MRQ both with pure loss and with queues.

Chapter 4
Models of Multi-Rate Systems with Inelastic Calls

As was mentioned Introduction, Part II, the operation of multi-service communication networks that handle heterogeneous traffic is described with fair accuracy by models of multi-rate queuing systems. In this chapter effective algorithms for calculation of QoS (Quality of Service) metrics of MRQs (Multi-Rate Queue) with inelastic calls are developed. These algorithms are used for general MRQ models with pure loss (i.e. without queues), in which the three most popular CAC (call admission control) are used: a simple complete sharing strategy, complete sharing strategy with equalization, and a strategy based on reservation of channels. It is shown that the developed algorithms have considerably lower complexity than appropriate counterparts.

We also describe here a method for calculation of QoS metrics of MRQ where two types of inelastic calls – narrowband (n-calls) and wideband (w-calls) are handled. We will call such models Gimpelson models, since to our knowledge Gimpelson first started earnest research of such models. Considering peculiarities of such models, an access strategy is suggested where a Special Group of Channels (SGC) is dedicated to w-calls. Results of numerical experiments are given, which are carried out by using the developed algorithms. Thorough analysis of these experiments is also given.

4.1 General Models of Unbuffered Multi-Rate Systems

In this section we consider models of MRQ without buffers. A sufficiently general model from this class can be described as follows.

The unbuffered system has N, $N > 1$, identical and parallel channels (slots, basic bandwidth units etc. depending on specific technology). The call arrival process is a stationary Poisson process with mean rate Λ, where each newly arriving call requires b_i, $1 \leq b_i \leq N$, channels simultaneously with probability σ_i, where $0 < \sigma_I < 1$, where $i = 1,2,\ldots,K$, and $\sigma_1 + \mathrm{L} + \sigma_K = 1$. Assume that the number of channels required to serve a particular call becomes known when this call arrives in the system. Then the input of a system in fact consists of K-type Poisson flows with arrival rate $\lambda_i := \Lambda\sigma_i$, for flow i; calls of type i request b_i channels whose service

L. Ponomarenko et al., *Performance Analysis and Optimization of Multi-Traffic on Communication Networks*, DOI 10.1007/978-3-642-15458-4_4,

start and end times are simultaneous. The service time for calls of type i has exponential distribution with means μ_i^{-1}, $i = 1,2,\ldots,K$.

Below these models are considered with different call admission control strategies.

4.1.1 Complete Sharing Strategy

First we will consider CAC based on a *Complete Sharing* (CS) strategy. According to this strategy all channels are impartially shared between heterogeneous calls, i.e. handling calls of type i might be assigned any b_i free channels; it is blocked and lost with probability 1 if at this moment the number of free channels is less than b_i, $i = 1,2,\ldots,K$.

On the basis of the assumption described above, a K-dimensional Markov chain may be used to describe the system at equilibrium, i.e. states of the system at any time are described by vectors $\mathbf{n} = (n_1,\ldots, n_K)$, where n_i is the number of type i calls in the system. Then state space S is defined by the expression:

$$S := \{\mathbf{n} : n_i = 0,\ldots,[N/b_i], (\mathbf{n}, \mathbf{b}) \leq N\}, \tag{4.1}$$

where $\mathbf{b} = (b_1,\ldots, b_K)$, $[x]$ is the greatest integer less than or equal to x, and $(\mathbf{n}, \mathbf{b})$ is the dot product of vectors $\mathbf{n}$ and $\mathbf{b}$, i.e. $(\mathbf{n}, \mathbf{b}) := \sum_{i=1}^{K} n_i b_i$. The condition $(\mathbf{n}, \mathbf{b}) \leq N$ in (4.1) indicates that in any admissible state $\mathbf{n}$ the number of busy channels does not exceed the total number of channels.

Transition intensities between states of the given K-dimensional MC are calculated as follows:

$$q(\mathbf{n},\mathbf{n}') = \begin{cases} \lambda_i I(f(\mathbf{n}) \geq b_i), & \text{if } \mathbf{n}' = \mathbf{n} + \mathbf{e_i}, \\ n_i\mu_i, \quad \text{if } \mathbf{n}' = \mathbf{n} - \mathbf{e_i}, \\ 0 \quad \text{in other cases}, \end{cases} \tag{4.2}$$

where e_i is the ith orthogonal vector in K-dimensional Euclidean space, $i = 1, 2,\ldots,K$.

The system of balance equations (SBE) for stationary probabilities $p(\mathbf{n})$, $\mathbf{n}\in S$ has the following form:

$$\begin{aligned} &\left(\sum_{i=1}^{K} \lambda_i I(f(\mathbf{n}) \geq b_i) + \sum_{i=1}^{K} n_i\mu_i\right) p(\mathbf{n}) = \sum_{i=1}^{K} \lambda_i p(\mathbf{n} - \mathbf{e_i}) I(n_i > 0) \\ &\quad + \sum_{i=1}^{K} (n_i + 1)\mu_i p(\mathbf{n} + \mathbf{e_i}) I(f(\mathbf{n}) \geq b_i), \quad \mathbf{n} \in S; \end{aligned} \tag{4.3}$$

$$\sum_{n\in S} p(n) = 1, \tag{4.4}$$

where $f(\mathbf{n}) :=$ N-$(\mathbf{n}, \mathbf{b})$ denotes the number of free channels in state $\mathbf{n}$.

Equation (4.4) is called the normalizing condition. It is well known that the stationary distribution of the given MC has a multiplicative form:

$$p(\boldsymbol{n}) = G^{-1}(N, K) \prod_{i=1}^{K} \frac{\nu_i^{n_i}}{n_i!}, \tag{4.5}$$

where $\nu_i := \lambda_i/\mu_i$, $G(N, K) := \sum_{n \in S} \prod_{i=1}^{K} \frac{\nu_i^{n_i}}{n_i!}$ is called the normalizing constant over state space (4.1).

From (4.5) we conclude that $G^{-1}(N, K) = p(\mathbf{0})$ where $\mathbf{0}$ is the K-dimension null vector.

The stationary distribution of the given MC makes it possible to compute the system QoS metrics. So, the main QoS metrics are the stationary blocking probabilities of calls of type $i(\mathrm{PB}_i)$

$$\mathrm{PB}_i(\mathrm{CS}) := \sum_{f(n)=0}^{b_i - 1} p(\mathbf{n}), i = 1, \ldots, K. \tag{4.6}$$

Note 4.1. From (4.6) conclude that $\mathrm{PB}_i(\mathrm{CS}) = \mathrm{PB}_j(\mathrm{CS})$ if $b_i = b_j$ for any values of loading parameters, i.e. if calls of different types require the same number of channels then their blocking probabilities are the same also.

From (4.6) might be found other QoS metrics of the system, for example rate of serviced calls of each type, weighted sum of blocking probabilities etc. The main indicator of channel utilization is the average number of busy channels ($\tilde{N}(\mathrm{CS})$). This QoS metric is calculated as follows

$$\tilde{N}(\mathrm{CS}) := \sum_{n \in S_{\mathrm{CS}}} (\mathbf{n}, \mathbf{b})\, p\,(\mathbf{n}). \tag{4.7}$$

The multiplicative form (4.5) provides the only alternative for finding the stationary distribution as a solution of super-large balance equations (4.3) and (4.4). However, the application of (4.5) to find the stationary distribution and hence QoS metrics (4.6) and (4.7) runs into the difficult problem of calculating $G(N, K)$, because the dimension of state space (4.1) increases exponentially with the increase of N and K.

In order to eliminate the mentioned computational difficulties in the calculation of QoS metrics (4.6) and (4.7) at large values of N and K and critical values of loading parameters of heterogeneous traffic we propose a new algorithm below.

The technique is based on splitting of the state space (4.1), such that:

$$S = \bigcup_{r=0}^{N} S_r\,,\; S_r \bigcap S_{r'} = \varnothing\,,\; r \neq r'\,, \tag{4.8}$$

where $S_r := \{\mathbf{n} \in S : (\mathbf{n}, \mathbf{b}) = r\}$.

In splitting (4.8) the class of states S_r contains all microstates from S in which the number of busy channels equals r, $r = 0,1,\ldots,N$. Then all microstates included in the subset S_r are combined into the single merged state $<r>$, and the merge function $U: S \to \hat{S}$ is built up, where $\hat{S} := \{0, 1, \ldots, N\}$ and $U(\mathbf{n}) = <r>$ if $\mathbf{n} \in S_r$, $r = 0,1,\ldots,N$.

From definition of the merged model we conclude that its stationary distribution $\left\{\pi\left(<r>\right) : <r> \in \hat{S}\right\}$ might be expressed by the stationary distribution of the initial model:

$$\pi\left(<r>\right) = \sum_{n \in S_r} p(\mathbf{n}), \tag{4.9}$$

$$\sum_{r=0}^{N} \pi\left(<r>\right) = 1. \tag{4.10}$$

From (4.8) to (4.10) we find that

$$\pi\left(<0>\right) = p\left(\mathbf{0}\right) \text{ or } G^{-1}\left(N, K\right) = \pi\left(<0>\right) \tag{4.11}$$

The QoS metrics (4.6) and (4.7) are calculated from the stationary distribution $\left\{\pi\left(<r>\right) : <r> \in \hat{S}\right\}$ in the following way:

$$\mathrm{PB}_i\left(\mathrm{CS}\right) = \sum_{j=0}^{b_i-1} \pi(<N-j>), i = 1,\ldots,K; \tag{4.12}$$

$$\tilde{N}\left(\mathrm{CS}\right) = \sum_{r=1}^{N} r\pi(<r>). \tag{4.13}$$

The stationary distribution of the merged model satisfies the following system of equations [7]:

$$\sum_{i=1}^{K} v_i b_i \pi\left(<r-b_i>\right) = r\pi\left(<r>\right), r = 1,\ldots,N, \tag{4.14}$$

$$\sum_{r=0}^{N} \pi\left(<r>\right) = 1, \tag{4.15}$$

where $\pi(<x>) = 0$ if $x < 0$.

For effective solving of equations (4.14) and (4.15) we propose a new algorithm below. First all traffic that requires the same number of channels is merged into one flow, i.e. we define the sets

$$A\left(r\right) := \{i : \text{type } i \text{ calls which require } r \text{ channels}\}, r = 1,\ldots, N. \tag{4.16}$$

By using (4.16) we calculate merged loads:

$$\hat{\nu}_r = \begin{cases} \sum\limits_{i \in A(r)} \nu_i, & \text{if } A(r) \neq \varnothing, \\ 0, & \text{if } A(r) = \varnothing. \end{cases} \tag{4.17}$$

From (4.17) we conclude that the system of equations (4.14) and (4.15) has the following extended matrix:

$$\begin{pmatrix} \hat{\nu}_1 & -1 & 0 & \ldots & 0 & 0 & 0 \\ 2\hat{\nu}_2 & \hat{\nu}_1 & -2 & \ldots & 0 & 0 & 0 \\ \cdot & \cdot & \cdot & \cdot & \cdot & \cdot & \cdot \\ N\hat{\nu}_N & (N-1)\hat{\nu}_{N-1} & (N-2)\hat{\nu}_{N-2} & \ldots & \hat{\nu}_1 & -N & 0 \\ 1 & 1 & 1 & \ldots & 1 & 1 & 1 \end{pmatrix}$$

Thus

$$\pi(\langle r \rangle) = g_r \pi(\langle 0 \rangle), r = 1, \ldots, N, \tag{4.18}$$

where g_r is determined by 1-D recurrence formulae

$$g_0 := 1, \; g_r = \frac{1}{r} \sum_{i=1}^{r} i\hat{\nu}_i g_{r-i}, \; r = \overline{1, N}. \tag{4.19}$$

Therefore, calculation of QoS metrics (4.6) and (4.7) has led to the simple formulae:

$$\text{PB}_i\,(\text{CS}) = \left(\sum_{j=N-r+1}^{N} g_j \right) \Big/ \left(\sum_{j=0}^{N} g_j \right), i \in A(r); \tag{4.20}$$

$$\tilde{N}(\text{CS}) = \left(\sum_{i=1}^{N} i g_i \right) \Big/ \left(\sum_{i=0}^{N} g_i \right). \tag{4.21}$$

From (4.11), (4.18), and (4.19) we find that the normalizing constant is calculated by

$$G(N, K) = \sum_{r=0}^{N} g_r.$$

A major advantage of the developed algorithm is its low computational complexity. Indeed, unlike known similar algorithms, here the complexity of computing QoS metrics is independent of the number of heterogeneous traffic of type K and is estimated as $O(N)$. Such invariance is guaranteed due to the merging of traffic which requires the same number of channels [see formula (4.16)].

4.1.2 Complete Sharing with Equalization Strategy

In the previous subsection it was assumed that all call traffic uses channels equally. However, with such an access strategy it is expected that with an increase of the width of calls their loss probability also increases, i.e.

$$\text{if } b_i > b_j \text{ then } \mathrm{PB}_i(\mathrm{CS}) \geq \mathrm{PB}_j(\mathrm{CS})\,. \tag{4.22}$$

The validity of condition (4.22) results from (4.6). This leads to a problem of absolute fair handling of heterogeneous calls in MRQ in terms of equalization of their loss probabilities. Consequently, there is a necessity in the organization of such call handling where heterogeneous calls have equal chances, i.e. an access strategy where $\mathrm{PB}_i = \mathrm{PB}_j$ for each i, j. The simplest method to achieve this goal is to equalize the width of all calls. This access strategy is called *Complete Sharing with Equalization* (CSE) [4].

The mentioned CSE-strategy can be achieved in the following way: the newly arrived call is accepted if at the moment of arrival the number of free channels is equal to or higher than b, where $b = \max\limits_i\{b_i\}$.

The state of the system when this strategy is used, as well as in the case of the CS-strategy, is determined at any time with K-dimensional vector $\mathbf{n} = (n_1, \ldots, n_K)$, where n_i means number of calls of type i in the system, $i = 1, \ldots, K$. However, when using the CSE-strategy the maximum number of calls of type i in the system is $\left[\frac{N-b}{b_i}\right] + 1$, i.e. the state space of the model in this case is determined as:

$$S := \left\{\mathbf{n} : n_i = \overline{0, \left[\frac{N-b}{b_i}\right] + 1},\ (\mathbf{n}, \mathbf{b}) \leq N\right\}. \tag{4.23}$$

Elements of the generating matrix of the appropriate Markov chain with state space (4.23) when using the CSE-strategy are determined as follows:

$$q\left(\mathbf{n}, \mathbf{n}'\right) = \begin{cases} \lambda_i I(f(\mathbf{n}) \geq b), & \text{if } \mathbf{n}' = \mathbf{n} + \mathbf{e_i}, \\ n_i\mu_i, & \text{if } \mathbf{n}' = \mathbf{n} - \mathbf{e_i}, \\ 0 & \text{in other cases.} \end{cases} \tag{4.24}$$

Since the examined MC is finite-dimensional and all its states are communicating, it has a stationary mode upon any positive values of incoming traffic parameters. The system of balance equations for this MC is built with relations (4.24). It has the following form:

$$\left(\sum_{i=1}^{K} \lambda_i\, I(f(\mathbf{n}) \geq b) + \sum_{i=1}^{K} n_i\mu_i\right) p(\mathbf{n}) = \sum_{i=1}^{K} \lambda_i p\left(\mathbf{n} - \mathbf{e_i}\right)\ I(f(\mathbf{n} - \mathbf{e_i}) \geq b)$$
$$+ \sum_{i=1}^{K} \left(\mu_i + 1\right) \mu_i p\left(\mathbf{n} + \mathbf{e_i}\right) I(f(\mathbf{n}) \geq b)\,. \tag{4.25}$$

The appropriate normalizing conditions should be added to this SBE. The main QoS metric for heterogeneous calls – loss probability, is equal for all types of calls. Let us denote it as PB(CSE). It is determined through stationary distribution as follows:

$$\mathrm{PB\,(CSE)} := \sum_{f(n)=0}^{b-1} p(\mathbf{n}). \tag{4.26}$$

The average number of busy channels in this strategy is determined like (4.7). Unlike the model with the CS-strategy, models with the CSE-strategy have no multiplicative solution of SBE (4.25) [see formulae (4.5)]. This fact is derived from the irreversibility of the examined MC [11]. According to this property, the stationary distribution of MC has a multiplicative presentation if the following is true: if there is a transition from state $\mathbf{n}$ into state $\mathbf{n}'$, then there should be an opposite transition from state $\mathbf{n}'$ to state $\mathbf{n}$. Indeed, with the aid of (4.24) we deduce that there is a transition from state $\mathbf{n}+\mathbf{e_i}$ into state $\mathbf{n}$, however, the opposite transition exists only when $f(\mathbf{n}) \geq b$, i.e. there is no multiplicative solution of SBE (4.25) when using the CSE-strategy. This substantially complicates finding a solution for SBE (4.25) upon large dimensions (4.23) and hence complicates the calculation of QoS metrics of the model when using the CSE-strategy.

In order to overcome the above-mentioned difficulties an approach suggested in the previous subsection for the CS-strategy can be applied here as well.

We are examining space splitting of (4.23) which is similar to (4.8). Then an appropriate merging function is built. Considering (4.26) we have the following relation:

$$\mathrm{PB\,(CSE)} = \sum_{i=N-b+1}^{N} \pi(<i>). \tag{4.27}$$

The stationary distribution of a merged model meets the following requirements [4]:

$$\pi\,(<i>) = \frac{1}{i}\sum_{j=1}^{K} v_i b_i \pi\,(<i-b_j>)\,F_j\big(i-b_j\big),\quad i=1,\ldots,N; \tag{4.28}$$

$$\sum_{i=0}^{N} \pi(<i>) = 1, \tag{4.29}$$

where $\pi\,(<x>) = 0$ if $x< 0$,

$$F_j\big(i-b_j\big) = \begin{cases} 1, & \text{if } i-b_j \leq N-b, \\ 0 & \text{in other cases.} \end{cases} \tag{4.30}$$

On the basis of the set of equations (4.28), (4.29), and (4.30) and considering the results for the CS-strategy, the following formulae to calculate the QoS metrics of the CSE-strategy are developed:

$$\mathrm{PB(CSE)} = \left(\sum_{j=N-b+1}^{N} g_j \right) \Big/ \left(\sum_{j=0}^{N} g_j \right), \tag{4.31}$$

$$\tilde{N}(\mathrm{CSE}) = \left(\sum_{i=1}^{N} i g_i \right) \Big/ \left(\sum_{i=0}^{N} g_i \right), \tag{4.32}$$

where

$$g_0 := 1, \ \ g_l = \frac{1}{l} \sum_{i=1}^{l} i \hat{v}_i g_{l-i} F_i (l-i), \ \ l = \overline{1,N}. \tag{4.33}$$

Paper [4] suggests the following formulae for calculation of functions (4.30):

$$F_j^* \left(i - b_j \right) = \begin{cases} 1, \text{ if } i \leq N - b \\ 0 \ \text{ in other cases.} \end{cases} \tag{4.34}$$

However, if function (4.30) is determined with the aid of formula (4.34), then from (4.28) it turns out that $\pi(<i>) = 0$ for all $i > \mathrm{N} - b$. However, this is not true since $\pi(<i>) > 0$ for all $i = 0, 1, \ldots, N$.

4.1.3 Trunk Reservation Strategy

The strategies introduced in the previous subsections are not suitable for cases with very different QoS requirements for heterogeneous traffic; neither do they allow effective use of capacity of channels. In lieu of this more complicated access strategies are used.

One of these fancy strategies is *Trunk Reservation* (TR). In accordance with this strategy a call is accepted only when the number of free channels upon its acceptance is more than the requested number of channels. The extra number of channels requested by a call that are required to be free in order to start the servicing of a given type of call is called the trunk reservation parameter for that call type. The value of the reservation parameter may either be fixed or dependent on the call type.

The primary intent of reservation strategies is to protect wide-band calls, since with the CS-strategy such calls are blocked more often than narrow-band calls.

Here we suggest an effective algorithm for calculation of parameters of a general MRQ model with the TR-strategy, where reservation parameters depend on the type of call. It is also shown that such a strategy is a generalization of two previous strategies – CS and CSE.

The examined trunk reservation strategy is determined as follows. If at the arrival epoch of a type i call, the number of free channels are more than or equal to b_i+r_i where $0 \le r_i \le N - b_i$, then this call is accepted; otherwise, the newly arrived call is lost (blocked) with probability 1. The parameter r_i is called the trunk reservation parameter for the type i call, $i=1,2,\ldots,K$. We assume that if $b_i = b_j$, then $r_i = r_j$, i.e. calls that require the same number of channels have the same trunk reservation parameter.

Note 4.3. If $r_i = 0$ for all i then we have CAC based on the CS-strategy of access; if $r_i = b - b_i$ for all i, $i=1,2,\ldots,K$, where $b = \max\{b_i: i=1,2,\ldots,K\}$ then we have CAC based on the CSE-strategy of access.

Note that under this CAC the maximum number of calls of type i in state $\mathbf{n}$ is $[((N-r_i)/b_i)-1]+1$. Since the state space of the model with the TR-strategy is defined as

$$S := \left\{\mathbf{n} : n_i = 0, 1, \ldots, \left[\frac{N - r_i}{b_i} - 1\right] + 1,\ i = \overline{1,K}; (\mathbf{n}, \mathbf{b}) \le N\right\}, \tag{4.35}$$

the transition intensities between states of the given K-dimensional MC are calculated as follows:

$$q\left(\mathbf{n}, \mathbf{n}'\right) = \begin{cases} \lambda_i I\left(f\left(\mathbf{n}\right) \ge b_i + r_i\right), & \text{if } \mathbf{n}' = \mathbf{n} + \mathbf{e_i} \\ n_i \mu_i, \quad \text{if } \mathbf{n}' = \mathbf{n} + \mathbf{e_i} \\ 0 \quad \text{in other cases.} \end{cases} \tag{4.36}$$

From (4.36) we conclude that SBE in this case has the following form:

$$\left(\sum_{i=1}^{K} \lambda_i I(f(\mathbf{n}) \ge b_i + r_i) + \sum_{i=1}^{K} n_i \mu_i\right) p(\mathbf{n}) = \sum_{i=1}^{K} \lambda_i p\left(\mathbf{n} - \mathbf{e_i}\right) I(f(\mathbf{n} - \mathbf{e_i}) \ge b_i + r_i)$$
$$+ \sum_{i=1}^{K} (n_i + 1)\, \mu_i p\left(\mathbf{n} + \mathbf{e}_i\right) I(f(\mathbf{n}) \ge b_i + r_i)\,, \tag{4.37}$$

$$\sum_{n \in S_{\mathrm{TR}}} p(\mathbf{n}) = 1\,. \tag{4.38}$$

The blocking probability of type i calls, $\mathrm{PB}_i(\mathrm{TR})$ can be calculated from the stationary distribution as follows:

$$\mathrm{PB}_i\left(\mathrm{TR}\right) = \sum_{f(n)=0}^{b_i + r_i - 1} p\left(\mathbf{n}\right),\ i = \overline{1,K}. \tag{4.39}$$

Another QoS metric is calculated analogously to (4.7). The main problem with calculation of QoS metrics is to find $p(\mathbf{n})$, $\mathbf{n} \in S$, since unlike with the CS-strategy, in this case as with the CSE-strategy there is no multiplicative solution for stationary distribution. The latter complicates solution at large dimensions of state space (4.35).

That is why here another method, based on the approach described in Sects. 4.1.1 and 4.1.2, is suggested. Since a detailed description of this approach was given earlier we will leave out intermediate calculations.

In this case QoS metric (4.39) might be determined by the stationary distribution of the merged model as follows:

$$\mathrm{PB}_i\,(\mathrm{TR}) = \sum_{j=0}^{b_i+r_i-1} \pi\,(< N-j >),\ \ i=\overline{1,K}.$$

The stationary distribution of the merged model satisfies the following conditions [19]:

$$\pi\,(< i >) = \frac{1}{i}\sum_{j=1}^{K} v_j b_j \pi\,(< i-b_j >)\, G_j\left(i-b_j\right),\ \ i=1,\ldots,N,$$

$$\sum_{i=0}^{N} \pi\,(<i>) = 1,$$

where

$$G_i\,(j) = \begin{cases} 1, & \text{if } j \le N-b_i-r_i, \\ 0 & \text{otherwise.} \end{cases}$$

By using the approach described in Sects. 4.1.1 and 4.1.2 we conclude that for the given CAC the stationary distribution of the merged model can be calculated as follows:

$$\pi\,(<i>)\ = g_i \pi\,(<0>), i =\ 1,\ldots,N,$$

where g_i is determined by 1-D recurrence formulae:

$$g_0 := 1, g_i = \frac{1}{i}\sum_{j=1}^{i} j\hat{v}_j g_{i-j} G_j\,(i-j),\ \ i=1,\ldots,N.$$

Therefore, the desired QoS metrics can be calculated as follows:

$$\mathrm{PB}_i\,(\mathrm{TR}) = \left(\sum_{j=N-i-r_i+1}^{N} g_j\right) \Bigg/ \left(\sum_{j=0}^{N} g_j\right);$$

$$\tilde{N}\,(\mathrm{TR}) = \left(\sum_{i=1}^{N} i g_i\right) \Bigg/ \left(\sum_{j=0}^{N} g_j\right).$$

4.1.4 Numerical Results

To keep this subsection clear numerical experiments for Gimpelson models are given, assuming that a narrow-band call (n-call) is handled by one channel, whereas a wide-band call (w-call) requires simultaneously m channels, $1 < m \le N$.

Apparently, the following correlation is true for the CS-strategy at any values of load and structural parameters of the model [see also (4.6)]: $PB_w(CS) > PB_n(CS)$. With use of the algorithm described in Sect. 4.1.1 and after appropriate mathematical transformations we get the following expressions for QoS metrics of the model:

$$PB_n\,(CS) = g_N \Big/ \left(\sum_{r=0}^{N} g_r \right),$$

$$PB_w\,(CS) = \left(\sum_{r=N-m+1}^{N} g_r \right) \Big/ \left(\sum_{r=0}^{N} g_r \right),$$

$$\tilde{N}\,(CS) = \left(\sum_{r=1}^{N} r g_r \right) \Big/ \left(\sum_{r=0}^{N} g_r \right),$$

where

$$g_0 = 1,\ g_r = \begin{cases} \dfrac{\nu_n^r}{r!}, & \text{if } r = \overline{1, m-1}, \\ \dfrac{1}{r}\left(\dfrac{\nu_n^r}{r!} + r\nu_w \right), & \text{if } r = m, \\ \dfrac{1}{r!}\left(\nu_n \cdot g_{r-1} + m\nu_w g_{r-m} \right), & \text{if } r = \overline{m+1, N}, \end{cases}$$

$\nu_x := \lambda_x/\mu_x$, λ_x is the rate of x-calls and μ_x is the rate at which they are serviced, $x \in \{n, w\}$.

Part of the numerical experiments results are shown in Figs. 4.1, 4.2, 4.3, 4.5, and 4.6. Analysis of these results leads to the following conclusions:

- values for functions PB_n and PB_w calculated with initial data from [6] are fully compatible with those derived in our numerical experiments, however the new algorithm proposed herein has much greater simplicity than the procedure in [6];
- generally speaking at fixed input load functions PB_n and PB_w from ν_n (or ν_w) are not monotonic (see Figs. 4.1 and 4.2);
- at fixed N, ν_n, and ν_w function PB_w increases monotonically relative to argument m, however the same is not true for PB_n (see Fig. 4.3);
- at fixed input load and fixed m function PB_w decreases monotonically relative to argument N, but function PB_n is only monotonically decreasing within intervals $N \in [Rm, (R + 1)m)$, where R is a natural number, i.e. at points where $N = Rm$ function value PB_n sharply increases compared to the previous point $N = Rm - 1$. This behavior of function PB_n has the following explanation: at values N multiple

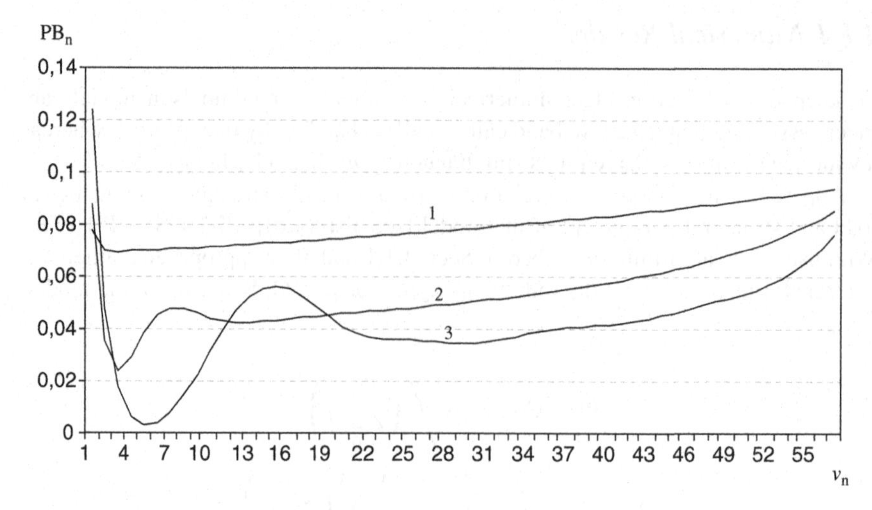

Fig. 4.1 PB_n versus v_n in CAC based on the CS-strategy, $N = 60$, $v_n + m\, v_w = 60$ Erl; $1 - m = 2$; $2 - m = 6$; $3 - m = 12$

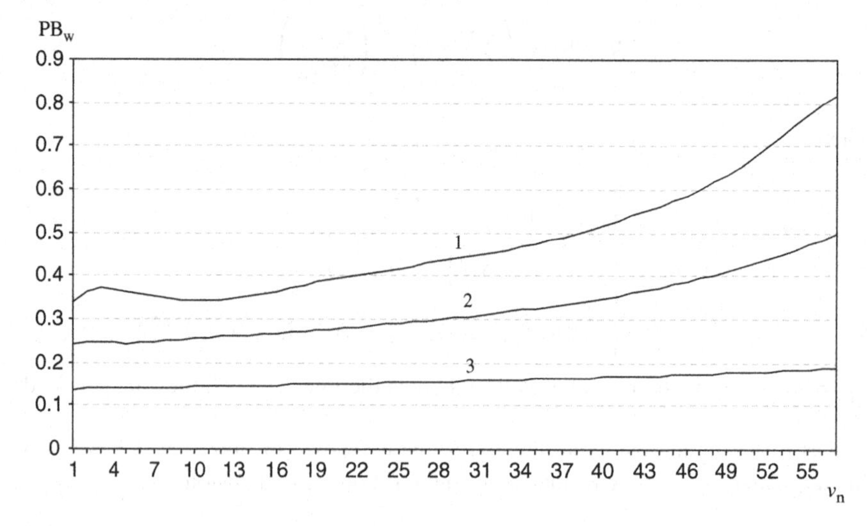

Fig. 4.2 PB_w versus v_n in CAC based on the CS-strategy, $N = 60$, $v_n + m\, v_w = 60$ Erl; $1 - m = 2$; $2 - m = 6$; $3 - m = 12$

of m, an additional possibility appears for w-calls to be accepted, hence chances for n-calls to be accepted worsen (see Fig. 4.4);

- function $\tilde{N}$ of m is not monotonic and is essentially dependent on values of input load parameters of the model (see Fig. 4.5);
- at any values of arrival intensities of different types of call and number of channels functions PB_n and PB_w decrease monotonically relative to handling intensities (see Fig. 4.6).

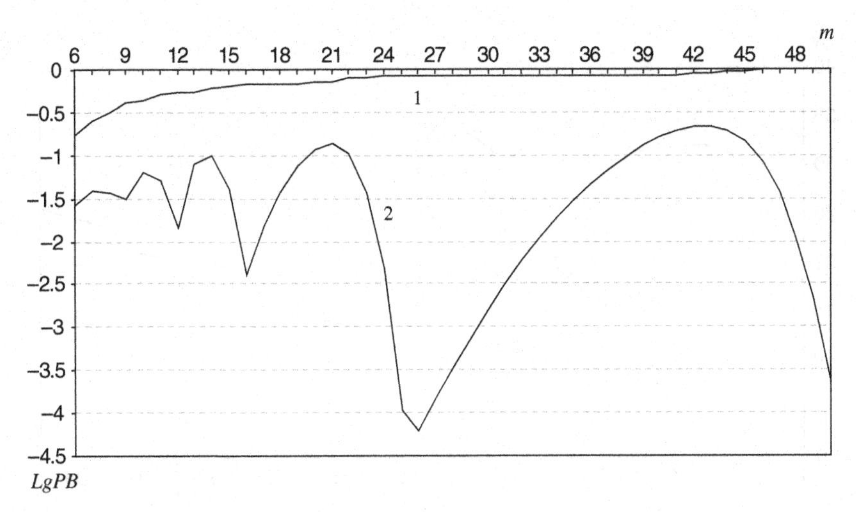

Fig. 4.3 Blocking probabilities versus m in CAC based on the CS-strategy, $N=50$, $v_n=10$ Erl, $v_w=5$ Erl; $1-PB_w$; $2-PB_n$

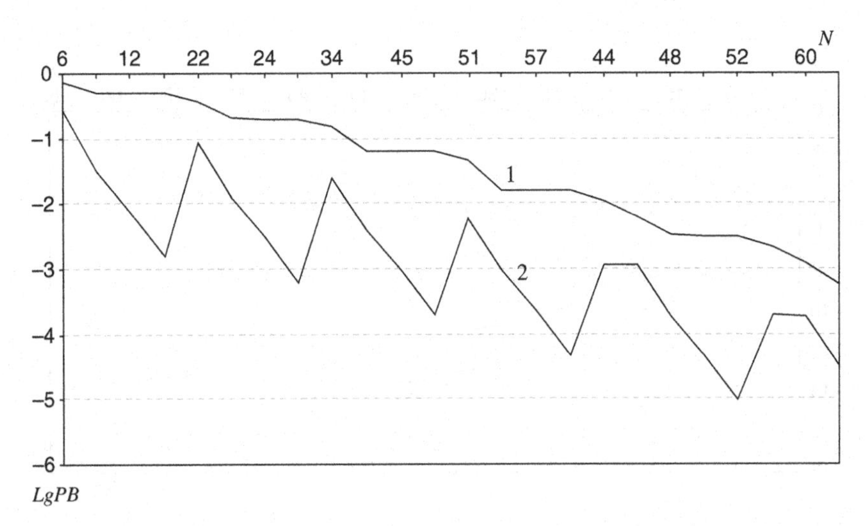

Fig. 4.4 Blocking probabilities versus N in CAC based on the CS-strategy, $m=6$, $v_n=v_w=1$ Erl; $1-PB_w$; $2-PB_n$

With the CSE-strategy in the Gimpelson model a newly arrived n-call or w-call is accepted only when the number of free channels equals or is higher than m, where m is the number of channels required for the w-call. For this strategy at any values of load and structural parameters we have:

$$PB_n(CSE) = PB_w(CSE) = PB(CSE).$$

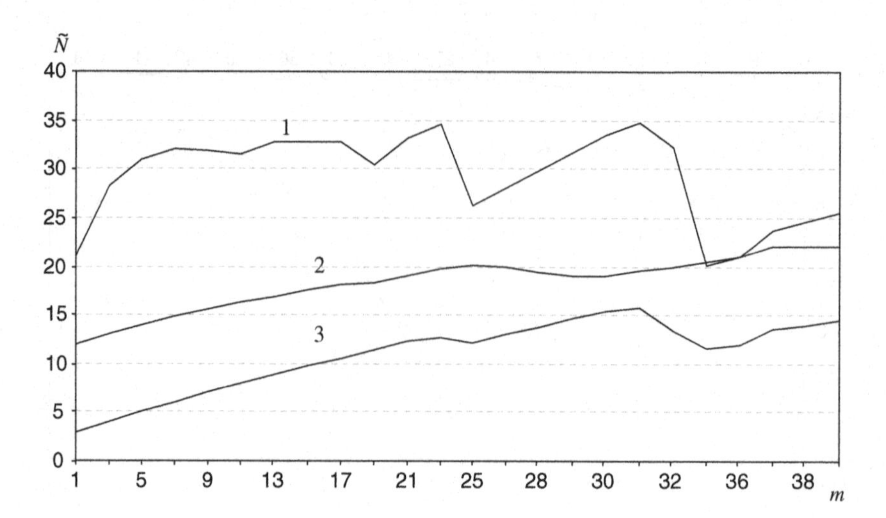

Fig. 4.5 $\tilde{N}$ versus m in CAC based on the CS-strategy, $N = 40$; $1 - \nu_n = 1$ Erl; $\nu_w = 10$ Erl; $2 - \nu_n = 10$ Erl; $\nu_w = 1$ Erl; $3 - \nu_n = 1$ Erl; $\nu_w = 1$ Erl

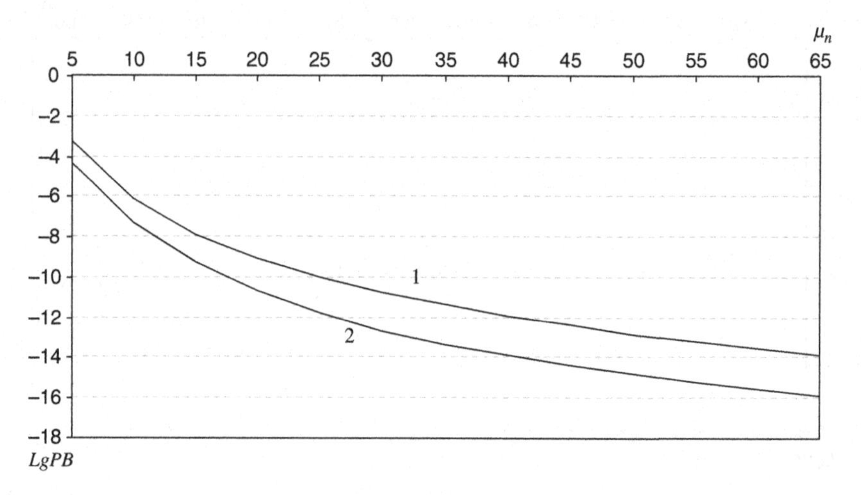

Fig. 4.6 Blocking probabilities versus μ_n in CAC based on the CS-strategy, $N = 60$, $m = 6$, $\lambda_n = \lambda_w = 50$; $1 - PB_w$; $2 - PB_n$

Using the algorithm developed in Sect. 4.1.2 and after some defined mathematical calculations it is possible to simplify to a great extent the procedures for calculation of $PB_n(CSE)$ [or $PB_w(CSE)$]. It is, however, essential to distinguish two different cases.

$$\text{Case } m \leq \left[\frac{N+1}{2}\right]:$$

$$g_r = \begin{cases} \dfrac{\nu_{\mathrm{n}}^r}{r!}, & \text{if } r = \overline{0, m-1}, \\ \dfrac{1}{r}\left(\nu_{\mathrm{n}} g_{r-1} I\,(m \le N - r + 1) + m\nu_{\mathrm{w}}\right), & \text{if } r = \overline{m, N}; \end{cases}$$

Case $m > \left[\dfrac{N+1}{2}\right]$:

$$g_r = \begin{cases} \dfrac{\nu_{\mathrm{n}}^r}{r!}, & \text{if } r = \overline{0, m-1}, \\ \dfrac{1}{r} m\nu_{\mathrm{w}} g_{r-m}, & \text{if } r = \overline{m, N}. \end{cases}$$

Part of the numerical experiments results are shown in Figs. 4.7, 4.8, 4.9, and 4.10. Analysis of these results leads to the following conclusions:

- with an increase of parameter m (i.e. with an increase of the width of w-calls) loss probabilities also increase which is an apparent result (see Fig. 4.7). However, at fixed total load (in our experiments $\nu_{\mathrm{n}} + \nu_{\mathrm{w}} = 12$ Erl) the rate of change of this function essentially depends on the ratio between traffic rates. At the same time the value of this function, starting from some value of parameter m, nears some limit (in our experiments this stated value of m is 18);
- the average number of busy channels is a decreasing function of parameter m, i.e. with an increase of the width of w-calls the utilization factor decreases (see Fig. 4.8). However, with a fixed total load the more n-calls are in the total

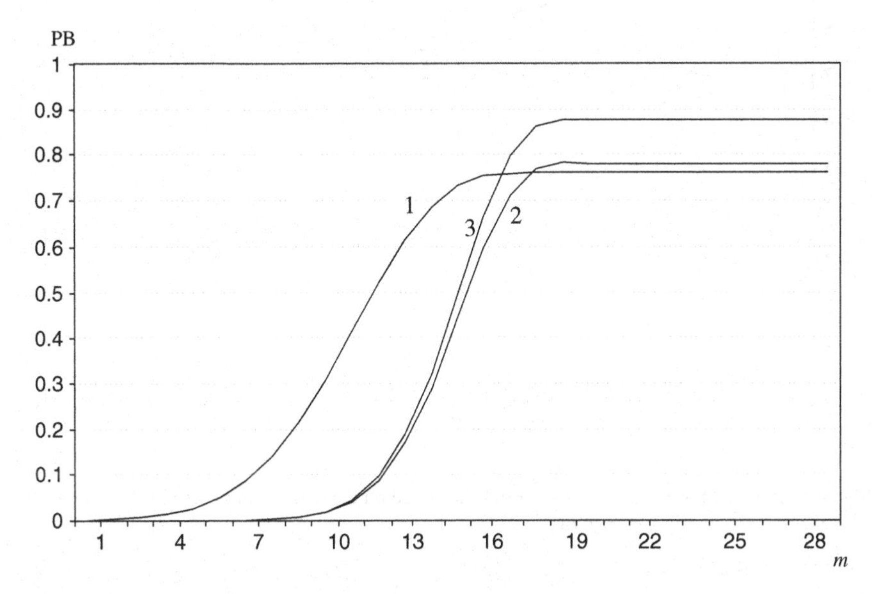

Fig. 4.7 Blocking probabilities versus m in CAC based on the CSE-strategy, $N = 50$; 1 – $\nu_{\mathrm{n}} = 8$ Erl, $\nu_{\mathrm{w}} = 4$ Erl; 2 – $\nu_{\mathrm{n}} = \nu_{\mathrm{w}} = 4$ Erl; 3 – $\nu_{\mathrm{n}} = 4$ Erl, $\nu_{\mathrm{w}} = 8$ Erl

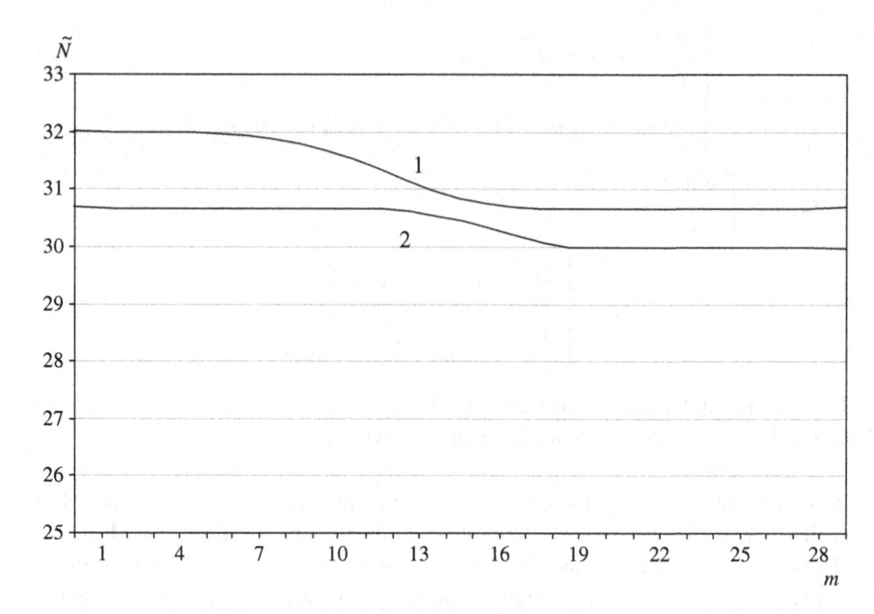

Fig. 4.8 $\tilde{N}$ versus m in CAC based on the CSE-strategy, $N = 50$; 1 – $\nu_n = 8$ Erl, $\nu_w = 4$ Erl; 2 – $\nu_n = 4$ Erl, $\nu_w = 8$ Erl

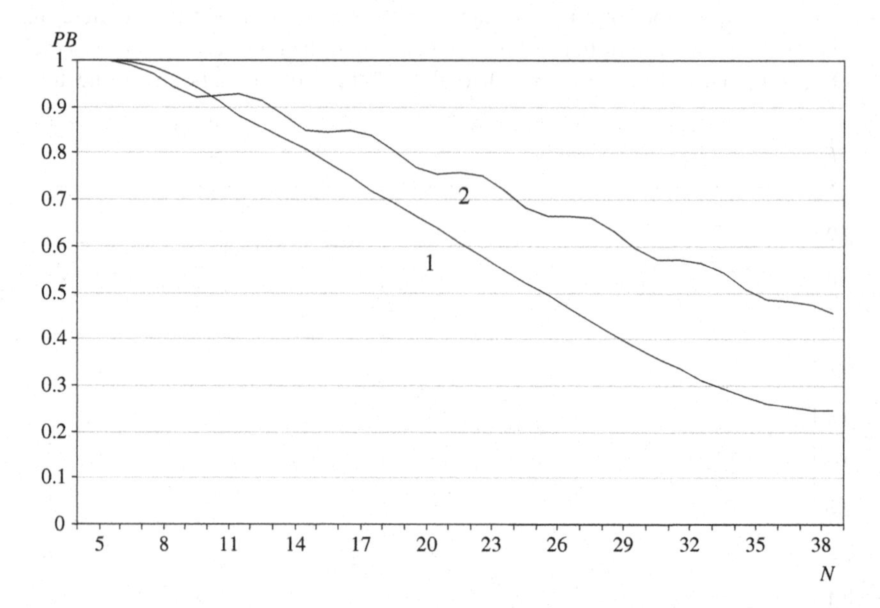

Fig. 4.9 Blocking probabilities versus N in CAC based on the CSE-strategy, $m = 5$; 1– $\nu_n = 8$ Erl, $\nu_w = 4$ Erl; 2– $\nu_n = 4$ Erl, $\nu_w = 8$ Erl

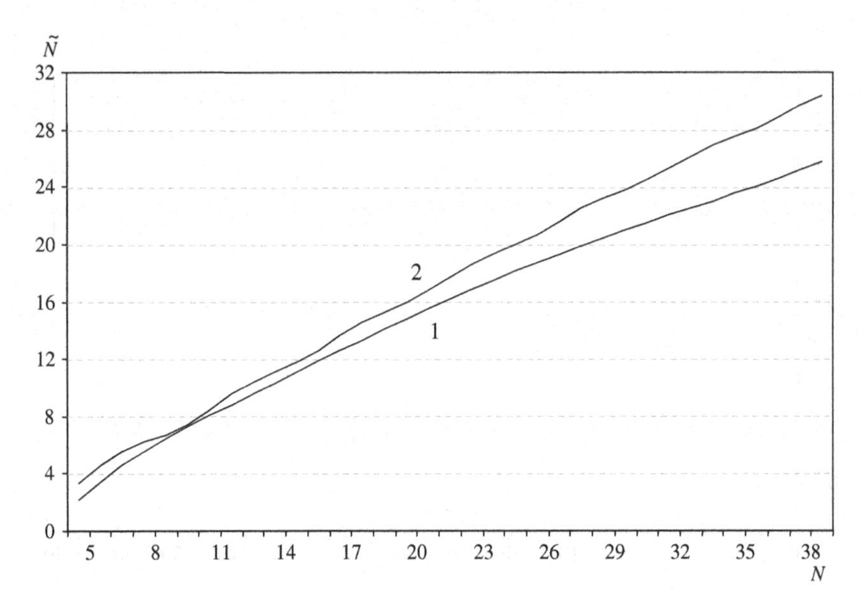

Fig. 4.10 $\tilde{N}$ versus N in CAC based on the CSE-strategy, $m = 5$; 1– $\nu_n = 8$ Erl, $\nu_w = 4$ Erl; 2– $\nu_n = 4$ Erl, $\nu_w = 8$ Erl

load, the greater the utilization factor. The latter fact also meets theoretical expectations;

- as was expected, loss probability is a decreasing function of argument N, therefore with a fixed total load the lower the number of w-calls in the total load, the lower the loss probability is (see Fig. 4.9). For instance, at $N = 25$ (m=5 is a fixed parameter) PB(4,8)/PB(8,4)$\approx$1.3, whereas at N=35, the same relation is approximately 2 (here PB(x, y) denotes loss probability at $\nu_n := x$ и $\nu_w := y$);
- the average number of busy channels is an increasing function of argument N (total number of channels), therefore with a fixed total load the higher the number of w-calls in the total load, the greater the utilization factor is (see Fig. 4.10).

Unlike the two previous models, for the TR-strategy it is not possible to find simple and explicit formulae for calculation of its QoS metrics. At the same time the appropriate algorithm is simple enough and allows analysis of QoS metrics in almost any range of load and structural parameters.

Dependency of model characteristics on reservation parameters r_n and r_m are plotted in Figs. 4.11, 4.12, 4.13, 4.14, 4.15, 4.16, 4.17, and 4.18. Other dependencies of these characteristics on load parameters ν_n and ν_m, as well as N and m, are not shown here due to space limitations in this book.

From Figs. 4.11 and 4.12 it can be seen that, at fixed values of r_n function PB_w increases with an increase of argument r_m whereas function PB_n decreases. These results are somewhat expected, since with the increase of a reservation parameter for w-calls their chances to be accepted decrease and at the same time chances for

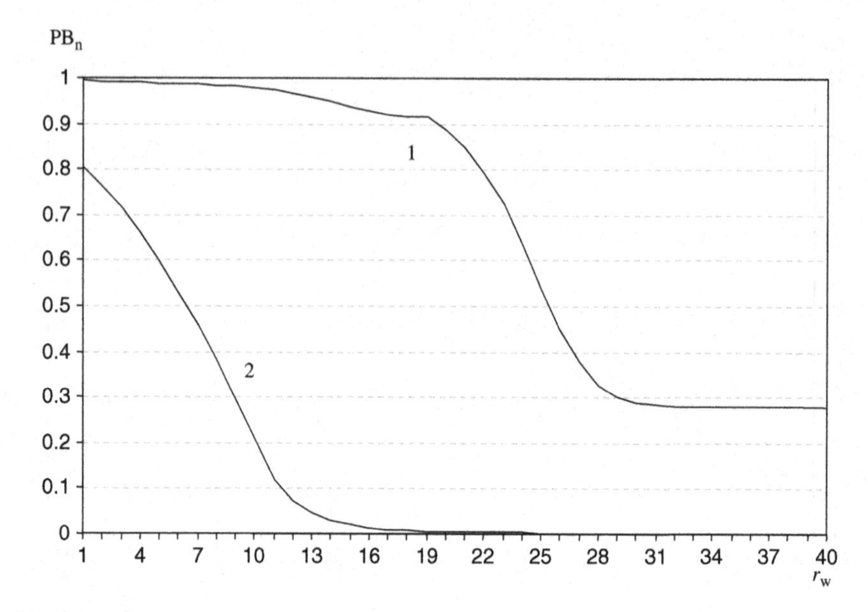

Fig. 4.11 PB_n versus r_w in CAC based on the TR-strategy, $N=50$, $m=10$, $v_n=25$ Erl, $v_w=20$ Erl; 1– $r_1=30$; 2– $r_1=10$

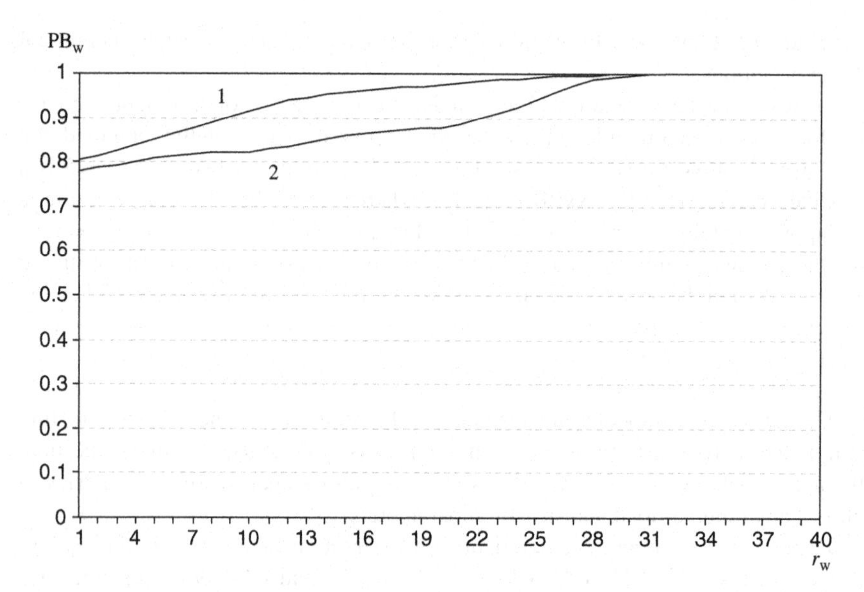

Fig. 4.12 PB_w versus r_w in CAC based on the TR-strategy, $N=50$, $m=10$, $v_n=25$ Erl, $v_w=20$ Erl; 1– $r_n=10$; 2– $r_n=30$

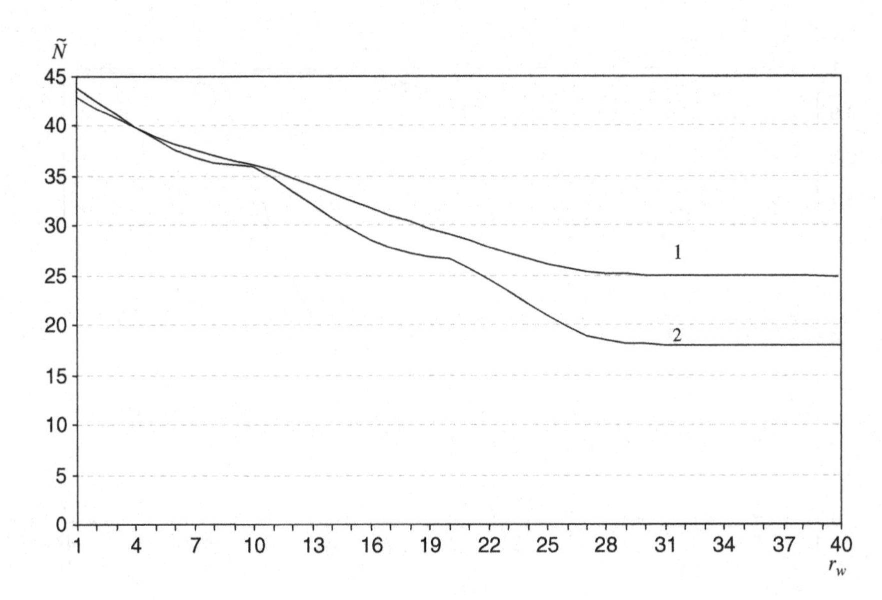

Fig. 4.13 $\tilde{N}$ versus r_w in CAC based on the TR-strategy, $N=50$, $m=10$, $v_n=25$ Erl, $v_w=20$ Erl; 1– $r_n=30$; 2– $r_n=10$

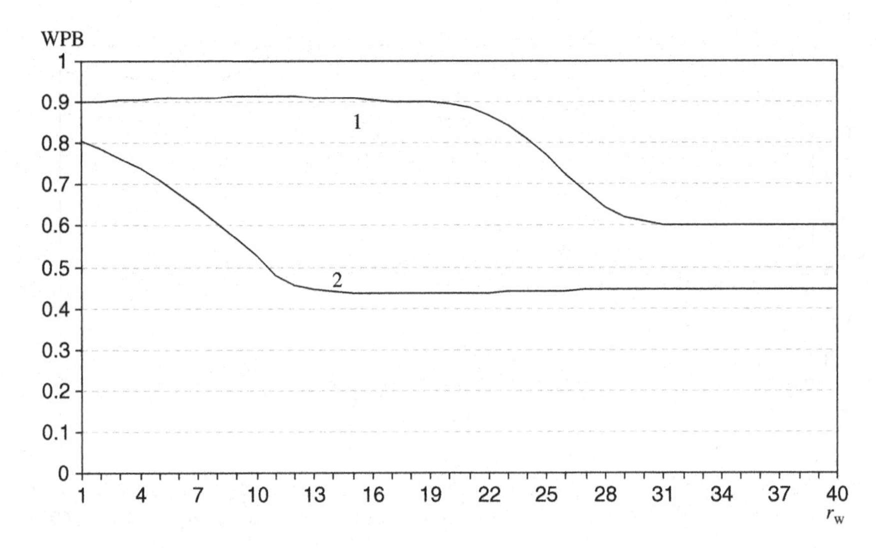

Fig. 4.14 Weighted sum of blocking probabilities versus r_w in CAC based on the TR-strategy, $N=50$, $m=10$, $v_n=25$ Erl, $v_w=20$ Erl; 1– $r_n=10$; 2– $r_n=30$

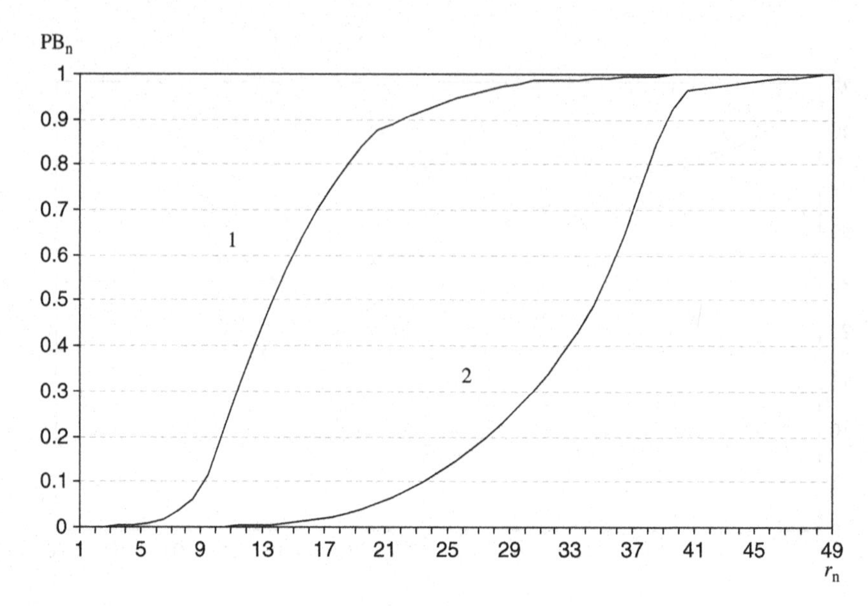

Fig. 4.15 PB$_n$ versus r_n in CAC based on the TR-strategy, $N=50$, $m=10$, $v_n=25$ Erl, $v_w=20$ Erl; 1– $r_w=10$; 2– $r_w=30$

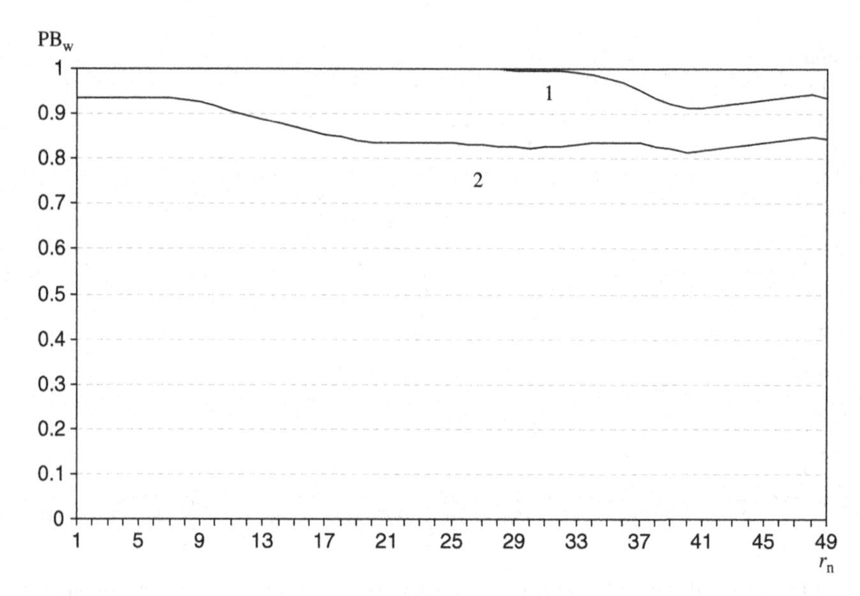

Fig. 4.16 PB$_w$ versus r_n in CAC based on the TR-strategy, $N=50$, $m=10$, $v_n=25$ Erl, $v_w=20$ Erl; 1 – $r_w=30$; 2– $r_w=10$

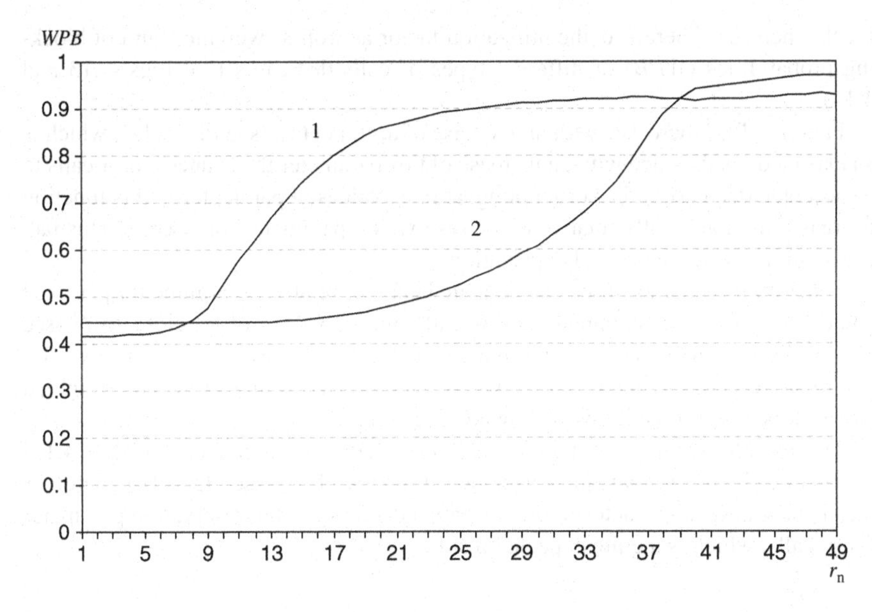

Fig. 4.17 Weighted sum of blocking probabilities versus r_n in CAC based on the TR-strategy, $N = 50$, $m = 10$, $v_n = 25$ Erl, $v_w = 20$ Erl; 1– $r_w = 10$; 2– $r_w = 30$

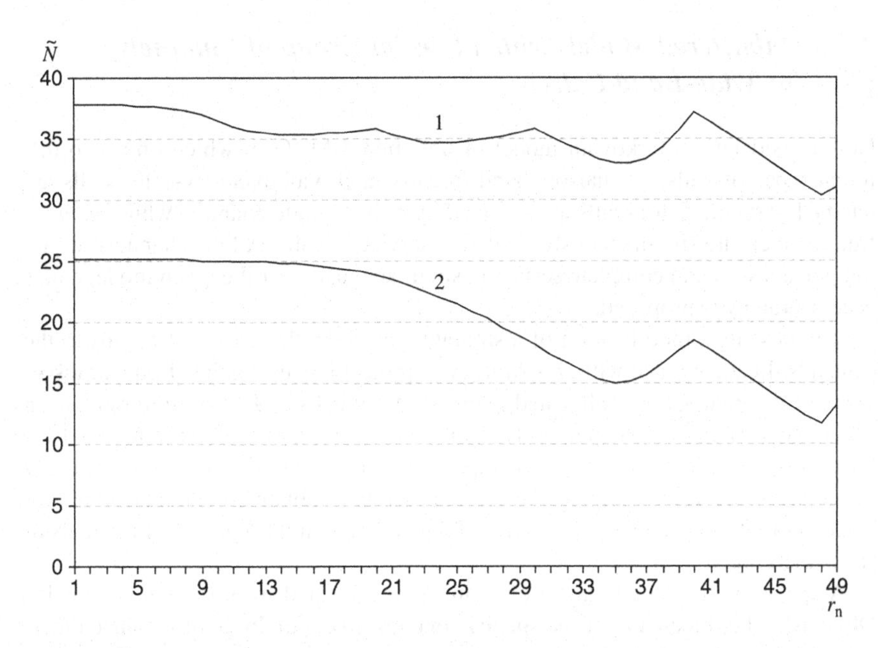

Fig. 4.18 $\tilde{N}$ versus r_n in CAC based on the TR-strategy, $N = 50$, $m = 10$, $v_n = 25$ Erl, $v_w = 20$ Erl; 1– $r_w = 10$; 2– $r_w = 30$

n-calls increase. Therefore, the utilization factor as well as weighted sum of blocking probabilities (*WPB*) of different types of calls decreases (see Figs. 4.13 and 4.14).

Function PB_n increases with an increase of argument r_n (see Fig. 4.15), which is an expected result, since with an increase of this parameter the chances for n-calls to be accepted decrease. It is worth noting, at fixed values of parameter r_n this function decreases monotonically relative to the reservation parameter for w-calls. This fact is also in line with theoretical expectations.

Function PB_w demonstrates inadequate behavior relative to argument r_n at fixed values of reservation parameter for w-calls and at values of r_n close to N (see Fig. 4.16). It is also worth noting that this function increases monotonically relative to the reservation parameter for w-calls and at fixed values of r_n. The weighted sum of blocking probabilities is shown in Fig. 4.17.

The non-monotonic behavior of function $\tilde{N}$ relative to argument r_n (see Fig. 4.18) is explained by the non-adequate behavior of function PB_w (see Fig. 4.16); therefore the channel utilization factor improves with a decrease of the reservation parameter for w-calls, which is a quite expected result.

4.2 Gimpelson-Type Multi-Rate Systems

4.2.1 Unbuffered Models with a Special Group of Channels for Wide-Band Calls

Now consider the Markovian model of unbuffered MRQ in which only two different types of calls, i.e. narrow-band (n-calls) and wide-band (w-calls) calls are handled. Assume that n-calls are serviced by only a single channel, while m, $m > 1$ free channels are simultaneously needed to service w-calls. All the channels servicing w-calls start and complete servicing simultaneously. For the given model a new access strategy is proposed.

It represents a modification of a strategy that limits the access of n-calls to the channels. In accordance with this strategy to protect w-calls against frequent losses some of the channels are allocated expressly to w-calls and a constraint either on the number of calls of a specific type or on both types of calls is introduced for the rest of the common zone. It is important to note that, despite the complexity of the proposed access strategy, here it is possible to obtain simple algorithms for calculation of the desired QoS metrics of a given system through use of methods of space merging.

A special part of the total set of channels is allocated for servicing of w-calls. Obviously, the dimension of the special zone must be a multiple of m, since otherwise this portion of the channels would be utilized inefficiently. Let us suppose that the dimension of the special zone is equal to mA, where A is a positive integer, $A = 1, 2, \ldots, \bar{A}$,

$$\bar{A} = \begin{cases} [N/m] - 1, & \text{if } \bmod(N, m) = 0, \\ [N/m] & \text{in other cases.} \end{cases} \tag{4.40}$$

Here and below, mod(x, y) denotes the remainder from the division of x by y. The pool for the rest of the channels of size N-mA is jointly used by both types of customers. However, since the formation of the special zone increases the chance of w-calls being accepted into the system, to compensate for the chance of n-calls a constraint is introduced on the number of w-calls in the common zone. In other words, it will be assumed that the number of w-calls in the common zone will not exceed some threshold value $\overline{k_{\mathrm{w}}}$: $\overline{k_{\mathrm{w}}} \le \left[\frac{N}{m} - A\right]$.

Note that if $\overline{k_{\mathrm{w}}} = \left[\frac{N}{m} - A\right]$, as a special case a strategy that limits the access of n-calls to the system with threshold parameter N-mA is obtained from the given strategy. Here we also note that if $\left[\frac{N}{m} - A\right] = 0$, the given strategy corresponds to a scheme of complete partitioning of the entire set of channels between the n- and w-calls. Therefore, it will henceforth be assumed that the parameter A is selected so that the condition $\overline{k_{\mathrm{w}}} \ge 1$ is satisfied.

Let us consider the problem of calculating the blocking probabilities of heterogeneous calls with the use of the proposed access strategy.

In the stationary mode the operation of a given system is described by a 2-D Markov chain with a state of the form $\mathbf{k} = (k_{\mathrm{n}}, k_{\mathrm{w}})$, where k_{n} (respectively, k_{w}) denotes the number of n-calls (respectively, w-calls) in the channels. Then according to the access strategy that has been described here, the state space of the given chain is defined thus:

$$S := \left\{\mathbf{k} : k_{\mathrm{n}} = 0, 1, \ldots, N - mA;\ k_{\mathrm{w}} = 0, 1, \ldots, k_{\mathrm{w}}^{\max}(k_{\mathrm{n}})\right\}, \tag{4.41}$$

where $k_{\mathrm{w}}^{\max}(k_{\mathrm{n}})$ denotes the maximally admissible number of w-calls in the system if the number of n-calls in the system is equal to k_{n}. It is easily proved that

$$k_{\mathrm{w}}^{\max}(k_{\mathrm{n}}) = A + \min\left\{\bar{k}_{\mathrm{w}}, \left[\frac{N - k_{\mathrm{n}}}{m} - A\right]\right\}. \tag{4.42}$$

In view of the operating principle of the proposed strategy, we conclude that the given Markov chain constitutes a model of a 2-D birth-and-death process (BDP) in the state space (4.40). The elements of its generating matrix $q(\mathbf{k}, \mathbf{k}')$, $\mathbf{k}, \mathbf{k}' \in S$ are determined thus:

$$q\left(\mathbf{k}, \mathbf{k}'\right) = \begin{cases} \lambda_{\mathrm{n}}, & \text{if } \mathbf{k}' = \mathbf{k} + \mathbf{e}_1, \\ \lambda_{\mathrm{w}}, & \text{if } \mathbf{k}' = \mathbf{k} + \mathbf{e}_2, \\ k_{\mathrm{n}}\mu_{\mathrm{n}}, & \text{if } \mathbf{k}' = \mathbf{k} - \mathbf{e}_1, \\ k_{\mathrm{w}}\mu_{\mathrm{w}}, & \text{if } \mathbf{k}' = \mathbf{k} - \mathbf{e}_2, \\ 0 & \text{in other cases.} \end{cases} \tag{4.43}$$

For the subsequent presentation we introduce the following representation of the state space (4.41):

$$S = \bigcup_{i=0}^{N-mA} S_i\,,\ S_i \bigcap S_j = \varnothing,\ i \ne j, \tag{4.44}$$

where $S_i := \{\mathbf{k} \in S : k_{\mathrm{n}} = i\}$.

Let us consider the problem of determining the cardinalities of the sets S_i in the decomposition (4.44). It is clear that sets S_i for $i = \overline{0, k_n^1}$, where $k_n^1 = N - m\left(A + \overline{k}_w\right)$ have maximum cardinalities. In other words, the cardinality of all sets S_i for $i = \overline{0, k_n^1}$ is equal to $A + \overline{k_w}$. Since decreasing the number of w-calls in the system by 1 makes it possible to increase the number of n-calls by precisely m units, it may be concluded that for $i = \overline{k_n^1 + (j-1)\,m + 1,\ k_n^1 + jm}$ the cardinality of the set S_i is given as $A + \overline{k_w} - j, j = 1, 2, \ldots, \overline{k_w}$.

The cardinalities of the sets S_i simultaneously determine the maximal possible number of w-calls in the given subset of states. Consequently, formula (4.42) may be rewritten thus:

$$k_w^{\max}(k_n) = \begin{cases} A + \overline{k}_w, & \text{if } 0 \le k_n \le k_n^1, \\ A + \overline{k}_w - i, & \text{if } k_n^1 + (i-1)\,m + 1 \le k_n \le k_n^1 + im,\ i = 1, 2, \ldots, \overline{k}_w. \end{cases} \tag{4.45}$$

The probabilities of a loss of n-calls (PB_n) and of w-calls (PB_w) are defined in terms of the stationary distribution $p(\mathbf{k})$, $\mathbf{k} \in S$ of the given Markov chain. In view of the operation of the proposed access strategy, the desired blocking probabilities are calculated thus:

$$PB_n := \sum_{i=0}^{\overline{k}_w} p\left(k_n^1 + im,\ A + \overline{k}_w - i\right) + \sum_{k \in S_{N-mA}} p(\mathbf{k}); \tag{4.46}$$

$$PB_w := \sum_{k_n=0}^{N-mA} p\left(k_n, k_m^{\max}(k_n)\right). \tag{4.47}$$

Let us consider an effective method which enables us to find the QoS metrics (4.46) and (4.47), and escape from the solving of large-scale balance equations for stationary distribution of the given model.

In the decomposition (4.44) the class of states S_i is described by a single merged state $<i>$ and the appropriate merged function is constructed (see Appendix). This function defines a merged (relative to the initial chain) Markov chain with state space $\hat{S} := \{< i >: i = 0, 1, \ldots, N - mA\}$.

The stationary distribution within the class S_i, $i = 0, \ldots, N - mA$, is defined as the distribution of a classical queuing system $M|M|k_w^{\max}(i)\,|0$ with load $\nu_w := \lambda_w/\mu_w$ Erl. After carrying out some algebra we obtain the following formulae for calculating the stationary distribution of the merged Markov model, which constitutes a model of a 1-D BDP:

$$\pi(< x >) = \begin{cases} \dfrac{\nu_n^x}{x!}\pi(< 0 >), & \text{if } 1 \le x \le k_n^1, \\ \dfrac{\nu_n^x}{x!}\prod\limits_{i=0}^{y_x}\left(1 - E_B\left(\nu_w, A + \overline{k}_w - i\right)\right)\pi(< 0 >), & \text{if} \\ k_n^1 + (j-1)\,m + 1 \le x \le k_n^1 + jm,\ j = \overline{1, \overline{k}_w}, & \end{cases} \tag{4.48}$$

where

$$\nu_n := \lambda_n/\mu_n,\ \pi(<0>) = \left(1 + \sum_{x=1}^{k_{\rm n}^1} \frac{\nu_{\rm n}^x}{x!} + \sum_{x=k_{\rm n}^1+1}^{k_{\rm n}^1+m\bar{k}_{\rm w}} \frac{\nu_{\rm n}^x}{x!} \prod_{i=0}^{y_x} \left(1 - E_{\rm B}\left(\nu_{\rm w}, A + \bar{k}_{\rm w} - i\right)\right)\right)^{-1}, \tag{4.49}$$

$$y_x = \begin{cases} \left[\dfrac{x - k_{\rm n}^1}{m}\right], & \text{if } \bmod\left(x - k_{\rm n}^1, m\right) = 0, \\ \left[\dfrac{x - k_{\rm n}^1}{m}\right] + 1, & \text{in other cases.} \end{cases}$$

Consequently, from (4.46) to (4.49) the following formulae are found to calculate the blocking probabilities of heterogeneous calls:

$$\mathrm{PB}_{\rm n} \approx \pi\left(< k_{\rm n}^1 + m\bar{k}_{\rm w} >\right) + \sum_{i=0}^{\bar{k}_{\rm w}-1} E_{\rm B}\left(\nu_{\rm w}, A + \bar{k}_{\rm w} - i\right) \pi\left(< k_{\rm n}^1 + im >\right); \tag{4.50}$$

$$\mathrm{PB}_{\rm w} \approx E_{\rm B}\left(\nu_{\rm w}, A + \bar{k}_{\rm w}\right) \sum_{i=0}^{k_{\rm n}^1} \pi\left(< i >\right) + \sum_{i=1}^{\bar{k}_{\rm w}} E_{\rm B}\left(\nu_{\rm w}, A + \bar{k}_{\rm w} - 1\right) \sum_{j=k_{\rm n}^1+(i-1)m+1}^{k_{\rm n}^1+im} \pi\left(< j >\right). \tag{4.51}$$

It is important to note the rather low complexity of the proposed algorithm for calculating the desired blocking probabilities, since formulae (4.50) and (4.51) make extensive use of Erlang's B-formula, which has even been tabulated

4.2.2 *Models with Guard Channels and Buffers for Wide-Band Calls*

In the previous sections models of MRQ with pure losses were investigated where to protect w-calls either shared reservation or isolated reservation of channels was used. Another preventive scheme for saving w-calls is organizing a buffer for waiting in a queue. Here a MRQ model with buffers for w-calls and a shared reservation scheme which is based on a guard channels scheme are proposed. Narrow-band calls are handled in accordance with the scheme with pure loss, i.e. non-accepted n-calls are blocked.

CAC for n-calls is defined as follows. If at the arriving epoch of an n-call the number of free channels is more than a given threshold then this call is accepted; otherwise it will be lost. As in Sect. 4.2.1, for effective use of channel capacity the value of the defined threshold should be a multiple of m, i.e. the newly arrived n-call is accepted if at this epoch the number of free channels is more than mG, $1 \le G \le \overline{G}$, where $\overline{G}$ is defined much as (4.40).

Handling of w-calls is performed as follows. If at the arriving epoch of a w-call the number of free channels is at least m then this call is accepted immediately; otherwise it will join the queue. After departure of the w-call the number of free channels becomes enough to handle one call of this kind since in these cases one w-call is chosen from the queue (if any). If after departure of an n-call the number of free channels becomes enough to handle one w-call then one w-call is chosen from the queue (if any); otherwise the released channel stands idle and w-calls continue to wait in the queue until enough free channels have been released. For the sake of brevity assume that in the queue of w-calls the FCFS (first-come-first-served) discipline is used. Here both models with limited and unlimited buffers for w-calls are considered.

First consider the model with a limited buffer. The state of the system at an arbitrary moment of time in stationary mode is described by 2-D vector $\mathbf{k} = (k_{\mathrm{n}}, k_{\mathrm{w}})$, where k_{n} (k_{w}) define the number of n-calls (w-calls) in the system. It is clear that

$$0 \le k_{\mathrm{n}} \le N - mG; \quad 0 \le k_{\mathrm{w}} \le \left[\frac{N}{m}\right] + R, \tag{4.52}$$

where R is the size of the buffer for waiting w-calls.

Considering the admitted CAC we conclude that if the system is in state $\mathbf{k} = (k_{\mathrm{n}}, k_{\mathrm{w}})$ then the number of w-call in channels $\left(k_{\mathrm{w}}^{\mathrm{s}}\right)$ and number of w-calls in the queue $\left(k_{\mathrm{w}}^{\mathrm{q}}\right)$ are calculated as follows:

$$k_{\mathrm{w}}^{\mathrm{s}} = \begin{cases} \left[\frac{N-k_{\mathrm{n}}}{m}\right], & \text{if } k_{\mathrm{w}} \ge \left[\frac{N-k_{\mathrm{n}}}{m}\right] \\ k_{\mathrm{w}} & \text{otherwise;} \end{cases}, \tag{4.53}$$

$$k_{\mathrm{w}}^{\mathrm{q}} = \begin{cases} k_{\mathrm{w}} - \left[\frac{N-k_{\mathrm{n}}}{m}\right], & \text{if } k_{\mathrm{w}} \ge \left[\frac{N-k_{\mathrm{n}}}{m}\right] \\ 0 & \text{otherwise.} \end{cases}, \tag{4.54}$$

From (4.52) to (4.54) we find that in each possible $\mathbf{k} = (k_{\mathrm{n}}, k_{\mathrm{w}})$ from state space S the following condition should hold: $0 \le k_{\mathrm{n}} + mk_{\mathrm{w}}^{\mathrm{s}} \le N$.

Elements of the generating matrix of the appropriate 2-D MC $q\left(\mathbf{k}, \mathbf{k}'\right)$, $\mathbf{k}, \mathbf{k}' \in S$ are calculated by

$$q\left(\mathbf{k}, \mathbf{k}'\right) = \begin{cases} \lambda_{\mathrm{n}} & \text{if } f(\mathbf{k}) > mG, \ \mathbf{k}' = \mathbf{k} + \mathbf{e}_1, \\ \lambda_{\mathrm{w}} & \text{if } \mathbf{k}' = \mathbf{k} + \mathbf{e}_2, \\ k_{\mathrm{n}}\mu_{\mathrm{n}} & \text{if } \mathbf{k}' = \mathbf{k} - \mathbf{e}_1, \\ k_{\mathrm{w}}^{\mathrm{s}}\mu_{\mathrm{w}} & \text{if } \mathbf{k}' = \mathbf{k} - \mathbf{e}_2, \\ 0 & \text{in other cases}, \end{cases} \tag{4.55}$$

where $f(\mathbf{k}) := N - k_{\mathrm{n}} - mk_{\mathrm{w}}^{\mathrm{s}}$ define the number of free channels in state $\mathbf{k} \in S$.

Major QoS metrics of the given models are probability of blocking of both types of calls. These quantities are calculated as follows:

$$\mathrm{PB_n} = \sum_{k \in S} p(\mathbf{k})\, I\left(N \le k_\mathrm{n} + m\left(k_\mathrm{w}^\mathrm{s} + G\right)\right), \tag{4.56}$$

$$\mathrm{PB_w} = \sum_{k \in S} p(\mathbf{k})\, \delta\left(k_\mathrm{w}^\mathrm{q}, R\right). \tag{4.57}$$

Here $\delta(i,j)$ denotes Kronecker's symbols: $\delta(i,j) = \begin{cases} 1 \text{ if } i = j, \\ 0 \text{ if } i \ne j. \end{cases}$

States probabilities satisfy the following balance equations and normalizing condition:

$$p(\mathbf{k})\left(\lambda_\mathrm{n} I\left(N > k_\mathrm{n} + m\left(k_\mathrm{w}^\mathrm{s} + A\right)\right) + \lambda_\mathrm{w} I\left(k_\mathrm{w} \le N + R - 1\right) + k_\mathrm{n}\mu_\mathrm{n} + k_\mathrm{w}^\mathrm{s}\mu_\mathrm{w}\right) =$$

$$\lambda_\mathrm{n} p(\mathbf{k} - \mathbf{e}_1) I\left(N > k_\mathrm{n} + m\left(k_\mathrm{w}^\mathrm{s} + A\right) + 1\right) + \lambda_\mathrm{w} p(\mathbf{k} - \mathbf{e}_2) I\left(k_\mathrm{w} > 0\right) +$$

$$(k_\mathrm{n} + 1)\mu_\mathrm{n} p(\mathbf{k} + \mathbf{e}_1) + \left(k_\mathrm{w}^\mathrm{s} + 1\right)\mu_\mathrm{w} p(\mathbf{k} + \mathbf{e}_2), \quad \mathbf{k} \in S; \tag{4.58}$$

$$\sum_{k \in S} p(\mathbf{k}) = 1. \tag{4.59}$$

Direct calculation of the stationary distribution from equations (4.58) and (4.59) is difficult for problems in large-scale models. Therefore, herein approximate calculation formulae are suggested.

The following splitting of state space S is considered:

$$S = \bigcup_{i=0}^{N-mG} S_i\,,\ S_i \bigcap S_j = \varnothing,\ i \ne j, \tag{4.60}$$

where $S_i := \{\mathbf{k} \in S : k_\mathrm{n} = i\},\ i = 0, 1, \ldots, N - mG$.

Stationary distributions of split models with state space S_i, $i = 0, 1, \ldots, \mathrm{N} - mG$ are calculated as follows:

$$\rho_i(j) = \begin{cases} \dfrac{{\nu_\mathrm{w}}^j}{j!}\rho_i(0), & \text{if } j = 1, \ldots, [b_i], \\ \left(\dfrac{\nu_\mathrm{w}}{[b_i]}\right)^j \cdot \dfrac{([b_i])^{b_i}}{[b_i]!}\rho_i(0), & \text{if } j = [b_i] + 1, \ldots, [b_i] + R, \end{cases} \tag{4.61}$$

where

$$b_i := \left[(\mathrm{N} - i)/m\right],\ \rho_i(0) = \left(\sum_{j=0}^{[b_i]} \frac{{\nu_\mathrm{w}}^j}{j!} + \frac{([b_i])^{b_i}}{[b_i]!}\sum_{j=[b_i]+1}^{[b_i]+R}\left(\frac{\nu_\mathrm{w}}{[b_i]}\right)^j\right)^{-1}. \tag{4.62}$$

Intensities between classes S_i and S_j which are denoted by $q(i, j)$ are calculated by

$$q(i,i+1)=\begin{cases}\lambda_n \sum_{j=0}^{[b_i-G]-1} \rho_i(j), & \text{if } \operatorname{mod}(N-i,m)\neq 0,\\ \lambda_n \sum_{j=0}^{[b_i-G]} \rho_i(j), & \text{if } \operatorname{mod}(N-i,m)=0;\end{cases} \tag{4.63}$$

$$q(i,i-1)=i\mu_n; \tag{4.64}$$

$$q(i,j)=0, \text{ if } |i-j|>1. \tag{4.65}$$

From (4.63) to (4.65) stationary probabilities of merged states (i.e. class of states S_i) which are denoted by $\pi(<i>)$ are calculated as follows:

$$\pi(<i>)=\frac{{\nu_n}^i}{i!}\prod_{j=1}^{i} q(j-1,j)\pi(<0>),\ i=1,\ldots,N-mG, \tag{4.66}$$

where

$$\pi(<0>)=\left(1+\sum_{i=1}^{N-mG}\frac{{\nu_n}^i}{i!}\prod_{j=1}^{i} q(j-1,j)\right)^{-1}. \tag{4.67}$$

Finally to calculate the desired QoS metrics the following approximate formulae are given:

$$\begin{aligned}PB_n\approx \sum_{i=0}^{N-mG}\Bigg(&\sum_{j=[b_i-G]}^{[b_i]+R} \rho_i(j)\,\pi(<i>)I(\operatorname{mod}(N-i,m)\neq 0)\\ &+\sum_{j=[b_i-G]+1}^{[b_i]} \rho_i(j)\,\pi(<i>)I(\operatorname{mod}(N-i,m)=0)\Bigg);\end{aligned} \tag{4.68}$$

$$PB_w\approx \sum_{i=0}^{N-mG} \rho_i([b_i]+R)\,\pi(<i>). \tag{4.69}$$

By using the proposed approach the MRQ models with infinite queues of w-calls might be investigated also. In this case the ergodicity condition in the split model with state space S_i is $\nu_w < [b_i]$. Since this condition should be fulfilled for any i for $i=0,1,\ldots,N-mG$, we find the ergodicity condition of these kinds of models: $\nu_w < G$. On fulfillment of the mentioned ergodicity condition the following algorithm might be proposed to calculate the blocking probability of n-calls.

Step 1. For $i = 0, 1, \ldots, N - mG$, calculate

$$\rho_i(j) = \begin{cases} \dfrac{{\nu_w}^j}{j!}\rho_i(0), & \text{if } j = 1, \ldots, [b_i], \\ \left(\dfrac{\nu_w}{[b_i]}\right)^j \cdot \dfrac{([b_i])^{[b_i]}}{[b_i]!}\rho_i(0), & \text{if } j \geq [b_i] + 1, \end{cases}$$

where

$$\rho_i(0) = \left(\sum_{j=0}^{[b_i]-1} \frac{{\nu_w}^j}{j!} + \frac{{\nu_w}^{[b_i]}}{([b_i]-1)!} \cdot \frac{1}{[b_i] - \nu_w}\right)^{-1}.$$

Step 2. Calculate the parameters $\pi(<i>)$ from formulae (4.66) and (4.67).
Step 3. Calculate

$$\mathrm{PB_n} \approx 1 - \sum_{i=0}^{N-mG} \left(\sum_{j=0}^{[b_i-G]-1} \rho_i(j)\,\pi(<i>)\,I(\mathrm{mod}(N-i,m) \neq 0) + \sum_{j=0}^{[b_i-G]} \rho_i(j)\,\pi(<i>)\,I(\mathrm{mod}(N-i,m) = 0) \right).$$

4.2.3 Numerical Results

Now let us analyze the numerical results of experiments conducted with the algorithms developed in Sects. 4.2.1 and 4.2.2. First consider the results for the model with the SGC-strategy. The objective of these experiments was to determine the behavior of $\mathrm{PB_n}$ and $\mathrm{PB_w}$ for different values of the parameters of the proposed access strategy (i.e., A and $\bar{k}_w$).

Some of the results of the numerical experiments are shown in Figs. 4.19 and 4.20. The following conclusions are deduced from an analysis of the figures:

- for fixed N, m, ν_n, and ν_w, the function $\mathrm{PB_n}$ is monotonically increasing and the function $\mathrm{PB_w}$, in contrast, monotonically decreasing with respect to both parameters A and $\bar{k}_w$;
- despite the fact that in individual intervals of short length the function $\mathrm{PB_w}$ is close to linear with respect to both arguments, over the entire range of definition it is nonlinear in nature. For example, in the closed interval [0, 10] the function $\mathrm{PB_w}(A)$ is approximated very well by the linear function $\mathrm{PB_w}(A) = -0.04A + 0.89$. However, at $A = 15$ the exact value $\mathrm{PB_w}(15) = 0.36$, while its approximate value, calculated according to the above linear expression, is equal to 0.29, and this divergence increases with increasing A. With respect to both

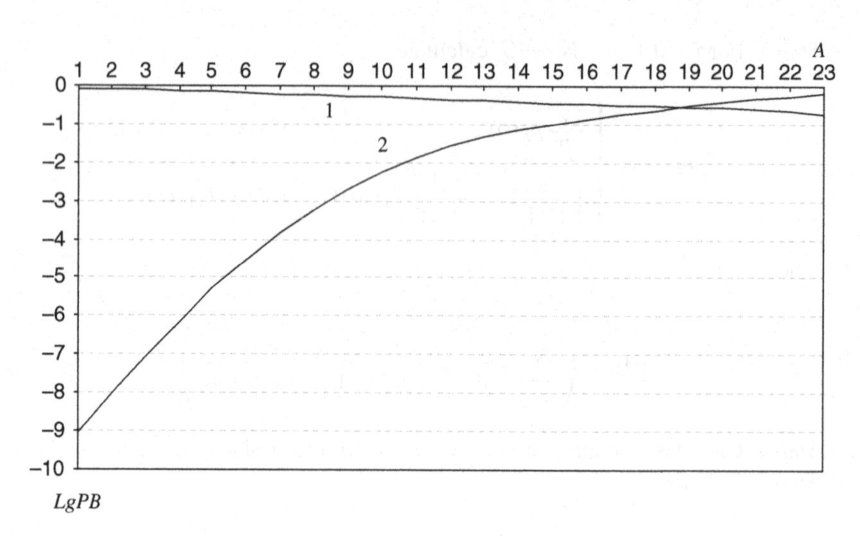

Fig. 4.19 Blocking probabilities versus A in CAC based on the SGC-strategy, $N=80$, $m=3$, $v_n=30$ Erl, $v_w=25$ Erl, $\overline{k_w}=3$; 1 – PB_w; 2 – PB_n

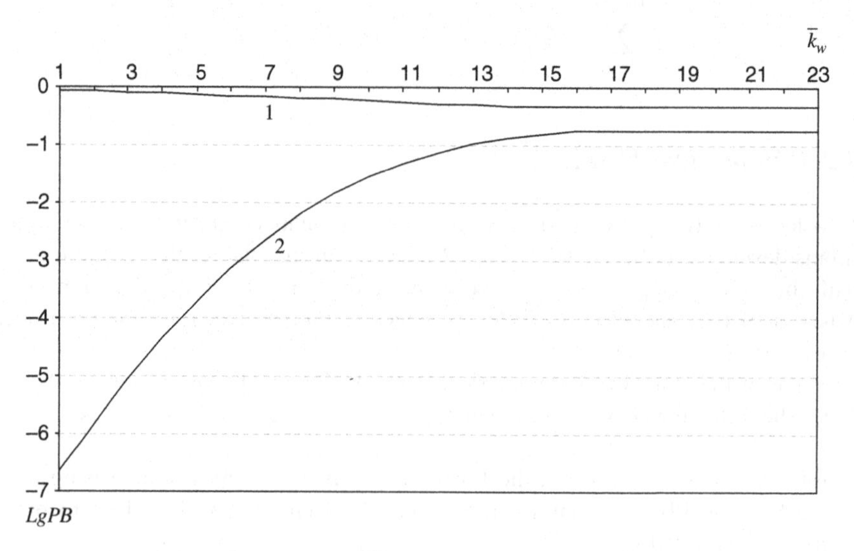

Fig. 4.20 Blocking probabilities versus $\overline{k_w}$ in CAC based on the SGC-strategy, $N=80$, $m=3$, $A=3$, $v_n=30$ Erl, $v_w=25$ Erl; 1 – PB_w; 2 – PB_n

arguments the function PB_n is nonlinear, and this property manifests itself particularly sharply with high values of A and low values of $\bar{k}_w$;

- the range of variation of the two functions PB_n and PB_w differ very strongly. Thus, the maximal value of $PB_w(A)$ (i.e., $PB_w(1)=0.85$) is nearly 4.5-times greater than the minimal value (i.e., $PB_w(23)=0.19$), and the corresponding

values $PB_n(23) = 0.65$ and $PB_n(1) = 0.1E\text{-}10$, i.e. $PB_w(23)/PB_w(1) \cong 10^{10}$. The same pattern occurs in studying the dependences of these functions on $\bar{k}_w$;
- the most fair servicing in the sense of maximal proximity of the values of $PB_n(A)$ and $PB_w(A)$ is attained with $A = 19$, in which case $PB_n(19) = 0.31$ and $PB_w(19) = 0.29$.

Finally some results of numerical experiments for the model with reserved channels and buffers for w-calls will be examined. Some results for models with limited buffers are given in Figs. 4.21, 4.22, 4.23, and 4.24. Analysis of these results leads to the following conclusions:

- at any permissible values of load and structural parameters of the system function PB_n increases monotonically relative to threshold parameter A (see Figs. 4.21 and 4.22). This is expected, as with the increase of threshold parameter A the chances for n-calls to be accepted decreases. In addition the increase of w-call load also leads to an increase of losses of n-calls (see Fig. 4.21), whereas an increase in w-call handling rate is accompanied by a decrease of loss intensity of n-calls (see Fig. 4.22);
- for any permissible values of load and structural parameters of the system function PB_w decreases monotonically relative to threshold parameter A (see Figs. 4.23 and 4.24). This is also expected, since with an increase of threshold parameter A the chances for w-calls to be accepted increases. Of interest is that for selected input data even a 10-fold increase of queue size for w-calls has

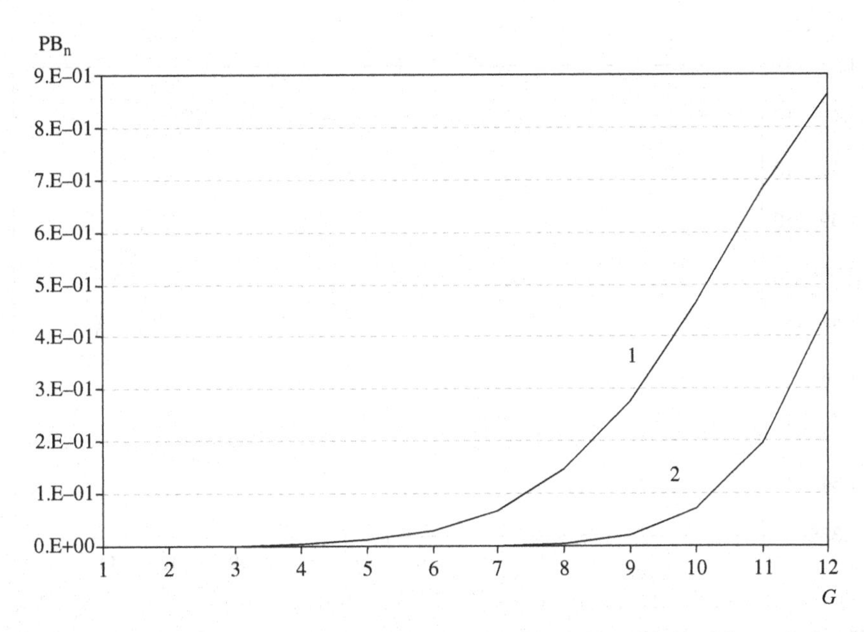

Fig. 4.21 PB_n versus G in the model with a limited buffer for w-calls, $N = 26$, $m = 2$, $R = 4$, $\lambda_n = 0.08$, $\mu_n = 0.5$, $\mu_w = 2$; $1 - \lambda_w = 7$, $2 - \lambda_w = 3$

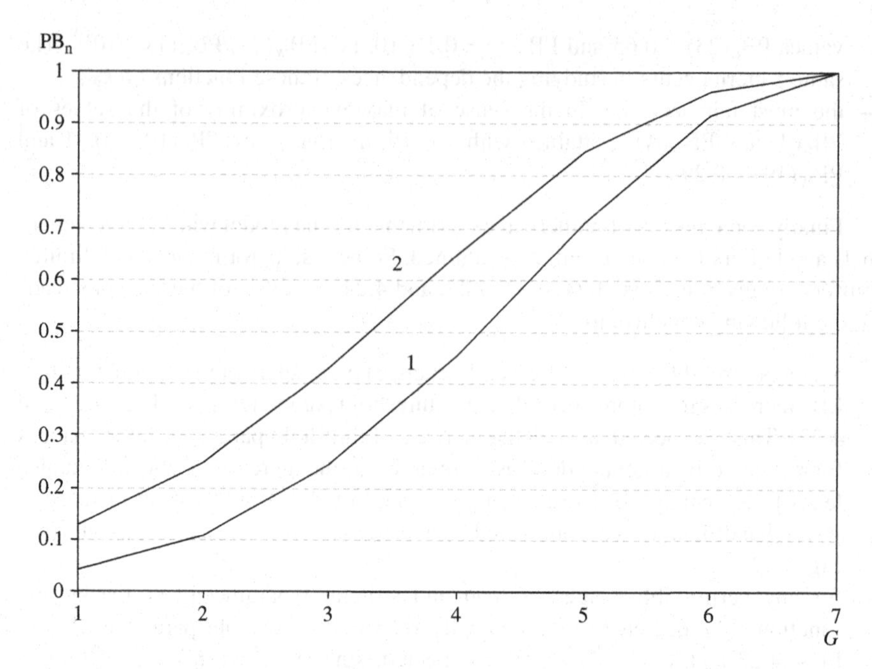

Fig. 4.22 PB_n versus G in the model with a limited buffer for w-calls, $N=15$, $m=2$, $R=4$, $\lambda_n=0.08$, $\lambda_w=10$, $\mu_n=0.5$; 1– $\mu_w=4$, 2– $\mu_w=3$

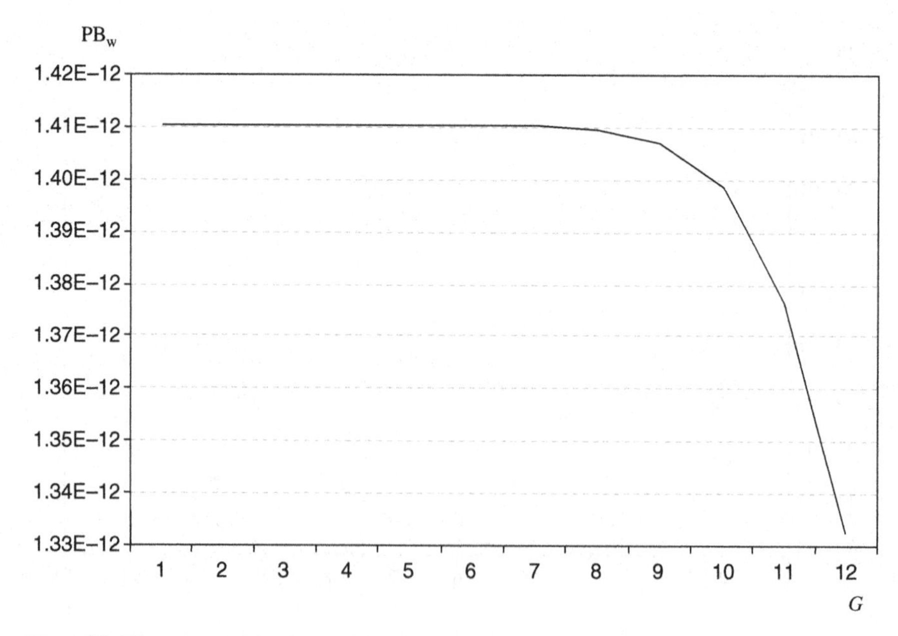

Fig. 4.23 PB_w versus G in the model with a limited buffer for w-calls, $N=26$, $m=2$, $R=4$, $\lambda_n=0.08$, $\lambda_w=3$, $\mu_n=0.5$, $\mu_w=2$

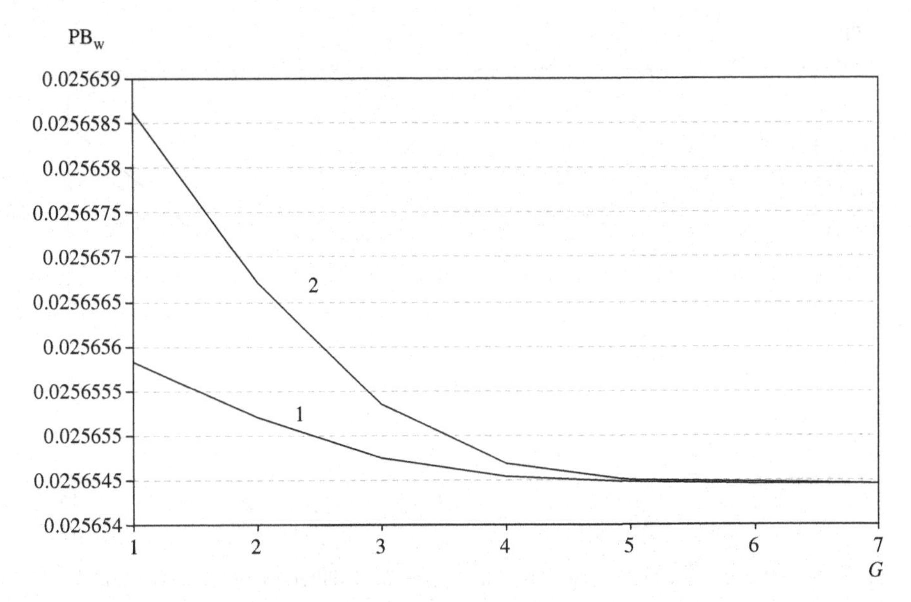

Fig. 4.24 PB_w versus G in the model with a limited buffer for w-calls, $N=15$, $m=2$, $R=4$, $\lambda_n=0.08$, $\lambda_w=10$, $\mu_w=2$; $1-\mu_n=0.7$, $2-\mu_n=0.4$

almost no influence on loss probability which is why Fig. 4.23 denotes only one curve of PB_w from A. This behavior for selected input data is explained by very low load of n-calls (see Fig. 4.23). Obviously, for other input data, sufficient dependency of function PB_w on queue size is expected. It is worth noting that an increase of handling rate of n-calls leads to a decrease of their loss probability (see Fig. 4.24). The latter is explained by the fact that with an increase of this handling rate the chances for w-calls to be accepted increases.

Results of numerical experiments for a model with an unlimited buffer for w-calls are shown in Fig. 4.25. As was expected, for this model also an increase of threshold parameter A decreases the chances for n-calls to be accepted, i.e. function PB_n increases monotonically relative to parameter A. Here the ergodicity condition of the model is true at a value of parameter $A \geq 4$, and therefore on the x-axis an appropriate range for parameter A is shown. Comparative analysis of the curves in Fig. 4.21 and Fig. 4.25 reveal that for the selected input data values of function PB_n in models with limited and unlimited buffers almost do not differ. This similarity takes place not only at low values of buffer (e.g. $R=4$), but also at sufficiently large values of buffer in a model with a limited buffer for w-calls (e.g. $R=40$). For these experiments this is explained by the low values of load parameters for n-calls ($\lambda_n=0.08$, $\mu_n = 0.5$).

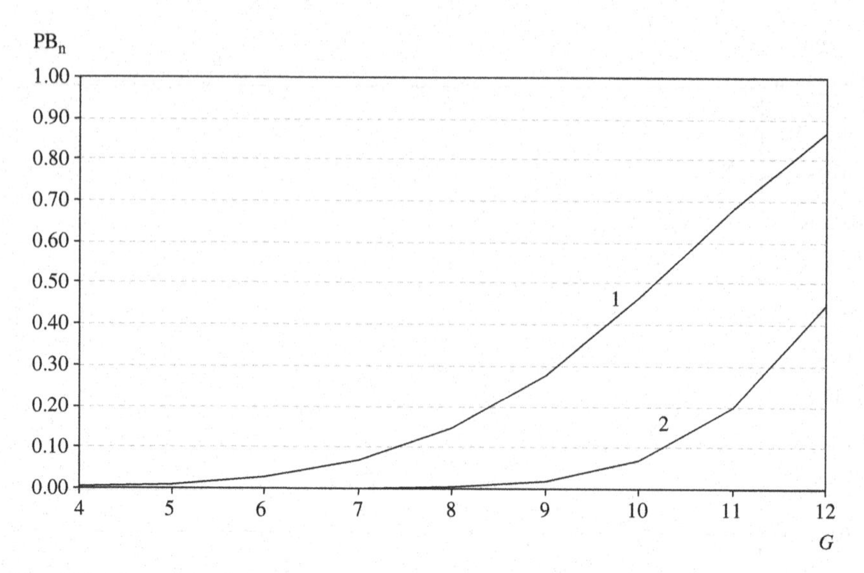

Fig. 4.25 PB_n versus G in the model with an unlimited buffer for w-calls, $N=26$, $m=2$, $\lambda_n=0.08$, $\mu_n=0.5$, $\mu_w=2$; $1-\lambda_w=7$, $2-\lambda_w=3$

4.3 Conclusion

A review of papers on MRQ as of 1995 can be found in [9, 11]. Analysis of works after this period fully support the suggestion made in [11], about the key role of research into MRQ models for application of queuing theory in teletraffic.

The most-studied models are MRQ models with non-elastic calls. Chronologically, L. Gimpelson's work [6] comes first. In this paper SBE is used to find the blocking probabilities for each type of calls, and for the numerical solution of this SBE it uses the iteration procedure of Gauss–Zeidel.

In [7, 22] it is shown that the stationary distribution of the general MRQ model has multiplicative form. According to [1] the multiplicative form exists also in cases when the handling time distribution function for different types of calls is arbitrary with a finite mean. It holds true as well for MRQ models of Engset type, when calls of different types arrive from finite sources.

Kaufman [7] and Roberts [22] almost simultaneously and independent from each other suggested an original approach for calculation of QoS metrics of multi-rate systems. Later on this algorithm is used in [10] for development of a more effective algorithm for calculation of QoS metrics. Recently in [25] a similar algorithm was suggested. It almost matches the one developed in [10].

MRQ models with more complex CAC are examined in [19, 26]. In [20] an MRQ model is researched where the CAC is based on a TR-strategy. By means of the Kaufman–Roberts algorithm, [20] suggests an approximate algorithm for QoS metrics calculation of a model with such a strategy. Then in [12] based on this algorithm an effective QoS calculation method is developed. An approximate algorithm for analyzing an MRQ model with CAC based on the SGC-strategy is proposed in [13].

In order to organize absolutely fair handling (in terms of equal blocking probabilities for different types of calls) in MRQ it is possible to use CAC based on a CSE-strategy. Such a model is examined in [4].

In the majority of works on MRQ, the considered models have fully switched resources (channels), i.e. they are said to handle i-calls with any b_i empty channels. At the same time, there is a rather vast class of MRQ, in which system channels may work only in some predefined sets due to either physical or economic limitations. Such models are investigated in [14, 23].

In some works MRQ models are studied with asymptotic methods and for this their asymptotic behavior is investigated at $N \to \infty$, $K \to \infty$, at critical input loads as well as for simultaneous increase of N, K, and load [2, 3, 5, 8, 15–18, 21, 24].

References

1. Burman DY, Lehoczky JP, Lim Y (1984) Insensitivity of blocking probabilities in a circuit-switching network. J Appl Prob 21(4):850–859
2. Choudhury GL, Leung KK, Whitt W (1995) An inversion algorithm to compute blocking probabilities in loss networks with state-dependent rates. IEEE/ACM Trans Netw 3(5):585–601
3. Chung S, Ross KW (1993) Reduced load approximations for multi-rate loss networks. IEEE/ACM Trans Netw 41(8):1222–1231
4. Delaire M, Hebuterne G (1997) Call blocking in multi-services systems on one transmission link. 5th international workshop on performance model and evaluation of ATM networks, July, UK, pp 253–270
5. Gazdzicki P, Lambadaris I, Mazumdar RR (1993) Blocking probabilities for large multi-rate Erlang loss system. Adv Appl Prob 25:997–1003
6. Gimpelson LA (1965) Analysis of mixtures of wide- and narrow-band traffic. IEEE Trans Commun Technol 13(3):258–266
7. Kaufman JS (1981) Blocking in shared resource environment. IEEE Trans Commun 10(10):1474–1481
8. Kelly FP (1986) Blocking probabilities in large circuit-switched networks. Adv Appl Prob 18:477–505
9. Kelly FP (1991) Loss networks. Ann Appl Prob 1(3):319–378
10. Melikov AZ (1993) An algorithm to calculation of multi-resource queues. Eng Simul 10(5):864–869
11. Melikov AZ (1996) Computation and optimization methods for multi-resource queues. Cybern Syst Anal 32(6):821–836
12. Melikov AZ, Deniz DZ (2000) Non-exhaustive channel access strategy in multi-resource communication systems with non-homogeneous traffic. In: Proceedings of 5th IEEE symposium on computers and communications, July 3–6, France, pp 432–437
13. Melikov AZ, Fattakhova MI, Kaziyev TS (2006) Multiple-speed system with specialized channels for servicing broadband customers. Autom Control Comput Sci 40(2):11–19
14. Melikov AZ, Molchanov AA, Ponomarenko LA (1993) Multi-resource queues with partially switchable channels. Eng Simul 10(2):370–379
15. Mitra D, Morrison JA (1994) Erlang capacity and uniform approximations for shared unbuffered resources. IEEE/ACM Trans Netw 2(6):558–570
16. Mitra D, Morrison JA, Ramakrishnan KG (1996) ATM Network design and optimization: A multirate loss network framework. IEEE/ACM Trans Netw 4(4):531–543

17. Mitra D, Morrison JA, Ramakrishnan KG (1999) Optimization and design of network routing refined asymptotic approximations. Perform Eval 36–37:267–288
18. Morrison JA (1994) Loss probabilities in a simple circuit-switched network. Adv Appl Prob 26:456–473
19. Omahen KJ (1977) Capacity bounds for multi-resource queues. J ACM 24(4):646–663
20. Pioro M, Lubacs J, Korner U (1990) Traffic engineering problems in multi-service circuit-switched networks. Comput Netw ISDN Syst 1–5:127–136
21. Puhalskii AA, Reiman MI (1998) A critically loaded multi-rate link with trunk reservation. Queuing Syst 28:157–190
22. Roberts JW (1981) A service system with heterogeneous user requirements application to multi-service telecommunications systems. In: Pujolle G (ed) Performance of data communication system and their applications. North Holland, Amsterdam, pp 423–431
23. Ross KW, Tsang DH (1990) Teletraffic engineering for product-form circuit-switched networks. Adv Appl Prob 22(3):657–675
24. Simonian A, Roberts JW, Theberge F, Mazumdar R (1997) Asymptotic estimates for blocking probabilities in a large multi-rate loss network. Adv Appl Prob 29:806–829
25. Taha S, Kavehrad M (2004) Dynamic bandwidth allocation in multi-class connection-oriented networks. Comput Commun 27:13–26
26. Whitt W (1985) Blocking when service is required from several facilities simultaneously. AT&T Tech J 64(8):1807–1856

Chapter 5
Models of Mixed Multi-Rate Systems

In this chapter models of multi-rate systems in which inelastic and elastic calls are jointly serviced are investigated. Elasticity of a call means that its width (i.e. the number of channels needed to service it) may vary as a function of the number of free channels. In such systems channel utilization may essentially be improved. Note that such models are not only of theoretical interest, but also of great interest for practical applications. In fact, in view of the development of the technology of coding and compression, some types of multimedia information in networks may be adapted to the available number of free bands. There are quite a number of such examples in telecommunication networks (cf. [3–5]).

Despite the widespread use of networks with inelastic and elastic messages, the mathematical models of such networks have not been subjected to sufficient study. There can be no doubt as to the critical importance of the latter type of models, since not all types of messages may become elastic.

Here an approach to calculate the models of MRQ with inelastic and elastic calls is proposed. The approach makes it possible to obtain efficient algorithms for calculating the QoS metrics of the particular models. In addition, models with both continuous and discrete bands for servicing elastic calls as well as models in which inelastic calls have pre-emptive-resume priority over elastic calls are investigated in detail.

In multi-service networks voice calls occupy a constant bandwidth during the lifetime of the connection while data calls receive a minimum bandwidth guarantee, and may be able to obtain more if the network is not overloaded. Hence in this chapter we will interpret inelastic calls as voice calls and elastic calls as the data variety.

5.1 Unbuffered Models

At first consider the following model of an N channel network which serves voice and data calls. The rate of Poisson traffic of v-calls (respectively, d-calls) is equal to λ_v (respectively, λ_d). The number of channels required to service a single v-call is a constant quantity and equal to b, $1 \leq b \leq N$, i.e. at least b free channels in the

L. Ponomarenko et al., *Performance Analysis and Optimization of Multi-Traffic on Communication Networks*, DOI 10.1007/978-3-642-15458-4_5,

system are needed to initiate servicing of a newly arrived v-call, and all the channels begin and end servicing of a given call simultaneously.

If there are m free channels, $\underline{m} \leq m \leq \overline{m}$ in the system at the moment a d-call arrives, the given call occupies all the free channels and begins to be serviced by all the channels simultaneously. After releasing channels, regardless of what type of call these channels had serviced, the channels switch to servicing d-calls. The service rate for v-calls has an exponential distribution with mean μ_{v}, while the service rate of d-calls is proportional to the number of busy channels and the service rate of the d-call if only one channel is used is μ_{d}.

In real multi-service networks d-calls may select their own service bands between $\underline{m}$ and $\overline{m}$ either continuously or discretely. Accordingly, these two types of models will be considered separately. For the sake of simplicity of presentation and without any loss in generality, it will henceforth be assumed that $b = 1$.

5.1.1 Models with a Continuous Band

We will assume that $\underline{m} = 1/N$, $\overline{m} = N$, i.e. the maximal number of d-calls that are being serviced is equal to N. If a d-call arrives when all the channels are free, it will occupy all of these channels and its servicing will proceed at a maximal rate. In general, d-calls occupy equally all those channels of the system that are not busy servicing v-calls, i.e. the total available channel capacity for them is equal to $(N - i)\mu_{\mathrm{d}}$ where i denotes the number of v-calls in the channels. In other words, a newly arrived d-call is rejected if at the moment it arrives all the system's channels are busy servicing v-calls, or if the number of d-calls in the system is equal to N. In precisely the same way, the maximum number of v-calls in the channels is equal to N (recall that each v-call requires only a single channel for its servicing). Moreover, if at the moment of arrival of a v-call there are no d-calls in the system and if there is at least one free channel, the given v-call is accepted for servicing; if at the moment of arrival of a v-call, there are N such calls already in the system, it is lost.

If at the moment of arrival of a v-call the number of such calls in the system is less than N–1 and if d-calls are being serviced, a single channel busy servicing d-calls is released for servicing the newly arrived v-call and, thus, the service rate for d-calls falls; if at the moment of arrival of a v-call, the number of such calls in the system is equal to $N - 1$ and if d-calls are being serviced (with overall rate μ_{d} since the bands are continuous and since all d-calls make equal use of the available bands), the newly arrived v-call is blocked.

The state of this system at an arbitrary moment of time may be described by a two-dimensional vector $\mathbf{k} = (k_1, k_2)$, where k_1 (respectively, k_2) denotes the number of voice (respectively, data) calls in the system. Then, the state space of the system is defined thus:

$$\mathrm{S} := \{\mathbf{k} : k_1 = 0, \ldots, N - 1, k_2 = 0, \ldots, N\} \cup \{(N, 0)\}. \tag{5.1}$$

On the basis of the mechanism of servicing of heterogeneous calls in this model, the possible transitions between states of the space (5.1) and their rates are determined in the following way. Upon the arrival of a v-call there occurs a transition from state $\mathbf{k}$ to state $\mathbf{k} + \mathbf{e_1}$ if $k_1 \leq N - 1$ and $k_2 = 0$ or $k_1 \leq N - 2$; upon the arrival of a d-call, there occurs a transition from state $\mathbf{k}$ to state $\mathbf{k} + \mathbf{e_2}$ if $k_1 < N-1$ and $k_2 \leq N-1$. The transition rate with the arrival of a v-call or a d-call is equal to λ_v or λ_d, respectively.

A transition from state k to state $\mathbf{k} - \mathbf{e_1}$ occurs at the moment a v-call exits the system and the rate of this transition is equal to $k_1\mu_v$; upon the departure of a d-call from the system in state $\mathbf{k}$ there occurs a transition into state $\mathbf{k} - \mathbf{e_2}$ and the rate of this transition is equal to $(N - k_1)\mu_d$.

Consequently, the elements of the generating matrix (GM) of the 2-D MC that describes the operation of the given MRQ are determined thus:

$$q\left(\mathbf{k},\mathbf{k}'\right) = \begin{cases} \lambda_v, & \text{if } k_1 \leq N-1,\ k_2 = 0 \text{ or } k_1 \leq N-2,\ \mathbf{k}' = \mathbf{k} + \mathbf{e_1}, \\ \lambda_d, & \text{if } k_1 \leq N-1,\ k_2 < N-1,\ \mathbf{k}' = \mathbf{k} + \mathbf{e_2}, \\ k_1\mu_v, & \text{if } \mathbf{k}' = \mathbf{k} - \mathbf{e_1}, \\ (N-k_1)\,\mu_d, & \text{if } \mathbf{k}' = \mathbf{k} - \mathbf{e_2}, \\ 0 & \text{in other cases.} \end{cases} \tag{5.2}$$

All the states of the present finite 2-D MC are communicated with each other; consequently, there exists a stationary distribution of the given Markov chain. The stationary probabilities of states $p(\mathbf{k})$, $\mathbf{k} \in S$, satisfy the following SBE, which is compiled on the basis of (5.2):

For state $\mathbf{k} = (N, 0)$:

$$N\mu_v p\left(\mathbf{k}\right) = \lambda_v p\left(\mathbf{k} - \mathbf{e_1}\right); \tag{5.3}$$

For states $\mathbf{k} \neq (N, 0)$:

$$\begin{aligned} (\lambda_v + \lambda_d + k_1\mu_v + (N-k_1)\,\mu_d)p\left(\mathbf{k}\right) &= \lambda_v p\left(\mathbf{k} - \mathbf{e_1}\right) + \lambda_d p\left(\mathbf{k} - \mathbf{e_2}\right) \\ &+ (k_1 + 1)\,\mu_v p\left(\mathbf{k} + \mathbf{e_1}\right) + (N-k_1)\,\mu_d p\left(\mathbf{k} + \mathbf{e_2}\right). \end{aligned} \tag{5.4}$$

The following normalizing condition is also added to the system of equations (5.3) and (5.4):

$$\sum_{k \in S} p\left(\mathbf{k}\right) = 1. \tag{5.5}$$

Equations (5.3), (5.4), and (5.5) are sufficient to find the stationary distribution in order to calculate the desired QoS metrics of the given model. Thus, the loss probabilities of v-calls (PB_v) and of d-calls (PB_d) are determined from the following relations:

$$\text{PB}_v = p\left(N, 0\right) + \sum_{i=1}^{N} p\left(N-1, i\right), \tag{5.6}$$

$$\mathrm{PB_d} = p(N,0) + \sum_{i=0}^{N-1} p(i,N). \tag{5.7}$$

The mean number of busy channels in this model is determined thus:

$$\tilde{N} = \sum_{i=1}^{N} ip(i,0) + N\left(1 - \sum_{i=0}^{N-1} p(i,0)\right). \tag{5.8}$$

Besides the metrics (5.6), (5.7), and (5.8), the probability of degradation of a d-call as well as the probability that the system is found in a state of degradation is no less important for this model. We now introduce the following concepts in order to define these metrics.

Since the d-calls make equal use of continuous bands, if the system is in state $\mathbf{k} = (i,j)$, the fraction of a band (respectively, service rate) for each d-call being serviced is equal to $(N-i)/j$ (respectively, $(N-i)\mu_d/j$), $j \neq 0$. However, in physical multimedia networks there exist constraints on the values of this parameter. In other words, if the rate of servicing of a d-call is less than some threshold value δ, $\delta>0$, the QoS is assumed to be unsatisfactory and, consequently, it is said that the system is in a state of degradation.

The probability of degradation states (PDS) is introduced in order to estimate this QoS metric. To calculate this quantity a subset of degradation states (S_{deg}) is determined in the state space (5.1):

$$S_{\mathrm{deg}} := \left\{\mathbf{k} \in S : \frac{N-k_1}{k_2} < \delta\right\}.$$

Then, based on the fact that the system is a Markovian system, we find the following formula for calculating the desired QoS metric:

$$\mathrm{PDS} = \sum_{k \in S_{\mathrm{deg}}} p(\mathbf{k}). \tag{5.9}$$

By means of (5.9) an alternative quantity to PDS may also be defined, i.e. the probability that the system is not in a state of degradation. In other words, the probability that d-calls are serviced with a required quality is equal to 1-PDS.

Another QoS metric, *the probability of d-call degradation* (PDD), is defined thus:

$$\mathrm{PDD} = M_{\mathrm{d}}^{\mathrm{deg}} / M_{\mathrm{d}}, \tag{5.10}$$

where M_{d} is the mean number of d-calls in the system and $M_{\mathrm{d}}{}^{\mathrm{deg}}$ the mean number of d-calls serviced at the degradation rate. The two quantities occurring on the right side of (5.10) are defined in a standard way (i.e. as the mean value of a discrete random variable):

$$M_{\rm d} = \sum_{i=1}^{N}\sum_{j=0}^{N-1} ip\,(j,i), M_{\rm d}^{\rm deg} = \sum_{i=1}^{N}\sum_{(i,j)\in S_{\rm deg}} ip\,(j,i)\,.$$

Thus, to calculate the QoS metrics (5.6), (5.7), (5.8), (5.9), and (5.10) it is necessary to find the stationary distribution of the given model. The solution of the latter problem does not present any special difficulty in the case of moderate dimensions of the state space (5.1). However, the solution of the problem encounters enormous computational difficulties where the state space (5.1) is of high dimension, since the SBE (5.3), (5.4), and (5.5) does not have an analytic solution.

An approach based on state space merging may be used to overcome these difficulties. In order to correctly apply this approach here it will be assumed that $\lambda_{\rm v}>>\lambda_{\rm d}$ (if the condition $\lambda_{\rm v}<<\lambda_{\rm d}$ holds, the procedure described below is modified by a well-known method).

With this assumption concerning the relationship between the loads of traffic of v- and d-calls, the value of the second component of the state vector is fixed and the following decomposition of the state space (5.1) is considered:

$$S = \bigcup_{i=0}^{N} S_i, \quad S_i \bigcap S_j = \varnothing,\ i \neq j, \tag{5.11}$$

where $S_i := \{\mathbf{k} \in S,\ k_2 = i\}, i = \overline{0,N}$.

The stationary distribution within class S_0 coincides with the well-known distribution of the classical Erlang's system *M/M/N/0* with load $v_{\rm v}$ Erl, i.e. $\theta_i(v_{\rm v}, N) = \frac{v_{\rm v}{}^i}{i!}\left(\sum_{j=0}^{N}\frac{v_{\rm v}{}^j}{j!}\right)^{-1}, i = \overline{0,N}$. At $k_2=j, j=1,\ldots,N$, the stationary distribution within all classes S_j are identical and coincide with the distribution of the states of the system *M/M/N–1/0* with load $v_{\rm v}$ Erl [see formula (5.2)].

Consequently, the transition rates between classes of microstates are determined from the relationships:

$$q(<i>,<j>) = \begin{cases} \lambda_{\rm d}\,(1 - E_{\rm B}\,(v_{\rm v}, N))\,, & \text{if } i = 0,\ j = i+1, \\ \lambda_{\rm d}, \quad \text{if } i > 0,\ j = i+1, & \\ \mu_{\rm d} \sum_{r=0}^{N-1} (N-r)\,\theta_r\,(v_{\rm v}, N-1), & \text{if } j = i-1, \\ 0 \quad \text{in other cases.} & \end{cases} \tag{5.12}$$

Then from (5.12) the stationary distribution of the merged model is calculated in an obvious way:

$$\pi\,(<j>) = \left(\prod_{i=0}^{j-1}\frac{q(<i>,<i+1>)}{q(<i+1>,<i>)}\right)\pi\,(<0>)\,, j = \overline{1,N}, \tag{5.13}$$

where

$$\pi(<0>) = \left(1 + \sum_{j=1}^{N} \prod_{i=0}^{j-1} \frac{q(<i>,<i+1>)}{q(<i+1>,<i>)}\right)^{-1}. \tag{5.14}$$

After some calculation, we find that the QoS metrics (5.6), (5.7), (5.8), (5.9), and (5.10) may be approximated from the following formulae:

$$\mathrm{PB}_v \approx E_\mathrm{B}(v_\mathrm{v}, N)\,\pi(<0>) + (N-1)\,E_\mathrm{B}(v_\mathrm{v}, N-1)\,; \tag{5.15}$$

$$\mathrm{PB_d} \approx E_\mathrm{B}(v_\mathrm{v}, N)\,\pi(<0>) + \pi(<N>)\,; \tag{5.16}$$

$$\tilde{N} \approx \sum_{i=1}^{N} i\theta_i(v_\mathrm{v}, N)\pi(<0>) + N(1 - \pi(<0>))\,; \tag{5.17}$$

$$\mathrm{PDS} \approx 1 - \sum_{i=0}^{N-1} \sum_{j=1}^{\left[\frac{N-i}{\delta}\right]} \pi(<j>)\,\theta_i(v_\mathrm{v}, N-1); \tag{5.18}$$

$$\mathrm{PDD} \approx M_\mathrm{d}^{\mathrm{deg}} / M_\mathrm{d}, \tag{5.19}$$

where

$$M_\mathrm{d} \approx \sum_{i=1}^{N} i\pi(<i>), M_\mathrm{d}^{\mathrm{deg}} \approx \sum_{i=1}^{N} \sum_{j=\left[\frac{N-i}{\delta}\right]+1}^{N-1} i\pi(<i>)\,\theta_j(v_\mathrm{v}, N-1)\,.$$

5.1.2 Models with a Discrete Band

Now let us consider a model in which d-calls may select bands in a discrete fashion, i.e. $\underline{m} = 1$, $\overline{m} = N$. As before (see Sect. 5.1.1), if a d-call arrives at a moment when the system is completely free, it will occupy all the channels; if a d-call occupies more than one channel, once a new d-call arrives, some of the channels that are busy servicing a d-call are switched over to service the newly arrived d-call. Moreover, any scheme of allocation of channels between d-calls that are being serviced may be used. Unlike a model with a continuous band, in the present case allocation of the capacity of the channels for a newly arrived d-call is not permitted if at this time all the channels in the system are busy.

The servicing of v-calls is performed in a similar way to what occurs in a model with a continuous band for d-calls (see Sect. 5.1.1). The state space of this model also contains 2-D vectors $\mathbf{k} = (k_1, k_2)$, but the state space of this model differs from (5.1) and is specified thus:

$$\mathrm{S} := \{\mathbf{k} : k_i = 0, \ldots, N, i = 1, 2; k_1 + k_2 \le N\}. \tag{5.20}$$

In view of the operating mechanism of this model we conclude that the elements of the GM corresponding to the given Markov chain are determined from the following relationships:

$$q\left(\mathbf{k},\mathbf{k}'\right)=\begin{cases}\lambda_{\mathrm{v}}, & \text{if}\,\mathbf{k}'=\mathbf{k}+\mathbf{e_1},\\ \lambda_{\mathrm{d}}, & \text{if}\,\mathbf{k}'=\mathbf{k}+\mathbf{e_2},\\ k_1\mu_{\mathrm{v}}, & \text{if}\,\mathbf{k}'=\mathbf{k}-\mathbf{e_1},\\ (N-k_1)\,\mu_{\mathrm{d}}, & \text{if}\,\mathbf{k}'=\mathbf{k}-\mathbf{e_2},\\ 0 & \text{in other cases.}\end{cases} \tag{5.21}$$

On the basis of (5.21) a system of balance equations for the stationary probabilities can be developed. It is analogous to (5.3), (5.4), and (5.5) so an explicit form of this system will not be presented here.

The QoS metrics of this model are determined by its stationary distribution in the following way. First, note that, unlike a model with a continuous band for d-calls, in the present model the blocking probabilities of heterogeneous calls are the same:

$$\mathrm{PB_v}=\mathrm{PB_d}=\sum_{k\in S_{\mathrm{d}}}p\left(\mathbf{k}\right), \tag{5.22}$$

where $S_{\mathrm{d}}:=\{\mathbf{k}\in S: k_1+k_2=N\}$ is a set of the diagonal states of the space (5.20).

The mean number of busy channels is determined by analogy with (5.8), though here it should be kept in mind that the stationary distribution of this model occurs on the right side of the mentioned formula.

For this model we also introduce the concept of the set of degradation states. However, the method of determining this set differs from the corresponding method for a model with a continuous band. Since fragmentation of each channel is not permitted in this model, it is assumed that the system is in a state of degradation if the total number of channels that are busy servicing d-calls is less than some predetermined threshold value $\delta>0$. In other words, state $\mathbf{k}$ is called a degradation state if for a given state, $k_1>N–\delta$ and $k_2>0$, where $1\leq\delta\leq N–1$. Thus, the probability of degradation states for the present model is determined thus:

$$\mathrm{PDS}:=\sum_{k\in S_{\mathrm{deg}}}p\left(\mathbf{k}\right), \tag{5.23}$$

where $S_{\mathrm{deg}}:=\{\mathbf{k}\in S: k_1\geq N-\delta,\, k_2>0\}$, $1\leq\delta\leq N-1$.

The probability of degradation of d-calls for the present model is determined by analogy with formulae (5.10).

The exact values of the QoS metrics are calculated by means of the stationary distribution of the model, which is determined by means of a SBE constructed on the basis of (5.23). As was noted earlier (see Sect. 5.1.1), with high dimensions of the state space of the model [in the present case, the state space determined from (5.20)] the process of finding exact values of the stationary probabilities of the states represents a rather complicated computational problem. Therefore, it is often necessary to limit the analysis to approximate values of these probabilities.

For the present model an approach considered in detail in Sect. 5.1.1 may also be used. Here we will present the final form of the algorithm for computing the QoS metrics of a model with a discrete band for d-calls.

Step 1. Compute $\pi\ (<i>)$, i=0,1,...,N, from (5.13), (5.14) where

$$q(<i>,<j>) = \begin{cases} \lambda_{\mathrm{d}}\left(1 - E_{\mathrm{B}}\left(\nu_{\mathrm{v}}, N-r\right)\right), & \text{if } j = i+1, \\ \mu_{\mathrm{d}} \sum\limits_{r=0}^{N-i} (N-r)\theta_r\left(\nu_{\mathrm{v}}, N-i\right), & \text{if } j = i-1, \\ 0 \quad \text{in other cases.} \end{cases}$$

Step 2. Compute approximate values of the QoS metrics of the model:

$$\mathrm{PB}_{\mathrm{v}} = \mathrm{PB}_{\mathrm{d}} \approx \sum_{i=0}^{N} E_{\mathrm{B}}\left(\nu_{\mathrm{v}}, N-i\right)\pi\left(<i>\right);$$

$$\tilde{N} \approx \pi\left(<0>\right)\sum_{i=1}^{N} i\theta_i\left(\nu_{\mathrm{v}}, N\right) + N\left(1 - \pi\left(<0>\right)\right);$$

$$\mathrm{PDS} \approx \sum_{i=N-\delta}^{N-1}\sum_{j=1}^{N-i} \pi\left(<j>\right)\theta_i\left(\nu_{\mathrm{v}}, N-j\right);$$

$$\mathrm{PDD} \approx M_{\mathrm{d}}^{\mathrm{deg}}/M_{\mathrm{d}},$$

where

$$M_{\mathrm{d}} = \sum_{i=1}^{N} i\pi\left(<i>\right),\ M_{\mathrm{d}}^{\mathrm{deg}} = \sum_{i=1}^{N}\sum_{j=N-\delta}^{N-i} i\pi\left(i\right)\theta_j\left(\nu_{\mathrm{v}}, N-i\right),$$

Thus, an efficient algorithm for computing approximate values of the QoS metrics of MRQ models with a discrete band for servicing d-calls has been developed. Like the corresponding algorithm for computing the QoS metrics of MRQ models with a continuous band for servicing d-calls, the present algorithm incorporates tabulated Erlangian values.

5.2 Models with Buffers for Elastic Calls

Now consider the model in which v-calls have pre-emptive-resume priority over d-calls and use only part of a channel, i.e. the number of v-calls in channels cannot exceed $m = [N/x]$ where x, $0 < x < 1$ denotes part of a single channel which is used by the v-call. As in Sect. 5.1.1, any scheme of fragmentation of each channel is allowed and it is assumed that v-calls are handled by the scheme: "blocking call is lost", i.e. a newly arrived v-call will be lost if there are m such calls in the channels already.

Calls of data will queue and unlike the previous models here a d-call at the head of the line will use all of the channels not currently occupied by v-calls. As in Sect. 5.1, the service rate of the d-calls is proportionate to the number of channels that it is using.

Here we study both cases: limited and unlimited buffers for d-calls. First consider the models with an unlimited buffer. The state of this system in stationary mode is described also by 2-D vectors $\mathbf{k}=(k_1,k_2)$ where k_1 is the number of v-calls in progress and k_2 is the number of d-calls in the system. Then state space S of the appropriate 2-D MC is

$$S = \{\mathbf{k} : k_1 = 0, 1, \ldots, m; k_2 = 0, 1, 2, \ldots\}. \tag{5.24}$$

The elements of the GM of this chain $q(\mathbf{k}, \mathbf{k}')$, $\mathbf{k}, \mathbf{k}' \in S$ are determined from the following relationships:

$$q(\mathbf{k}, \mathbf{k}') = \begin{cases} \lambda_{\mathrm{v}} & \text{if } \mathbf{k}' = \mathbf{k} + \mathbf{e_1} \\ \lambda_{\mathrm{d}} & \text{if } \mathbf{k}' = \mathbf{k} + \mathbf{e_2} \\ k_1 \mu_{\mathrm{v}} & \text{if } \mathbf{k}' = \mathbf{k} - \mathbf{e_1} \\ (N - x k_1)\, \mu_{\mathrm{d}} & \text{if } \mathbf{k}' = \mathbf{k} - \mathbf{e_2} \\ 0 & \text{otherwise.} \end{cases} \tag{5.25}$$

By using (5.25) the appropriate SBE for a given 2-D MC is constructed as follows:

$$\begin{aligned} (\lambda_v \mathrm{I}\,(k_1 < m) + \lambda_{\mathrm{d}} + k_1 \mu_v + (N - k_1 \cdot x)\mu_{\mathrm{d}})\, p\,(\mathbf{k}) = {} & \lambda_v p\,(\mathbf{k} - \mathbf{e_1})\, \mathrm{I}\,(k_1 > 0) \\ & + \lambda_{\mathrm{d}} p\,(\mathbf{k} - \mathbf{e_2})\, \mathrm{I}\,(k_2 > 0) + (k_1 + 1)\, \mu_v p\,(\mathbf{k} + \mathbf{e_1})\, \mathrm{I}(k_1 < m) \\ & + (N - k_1 \cdot x)\, \mu_{\mathrm{d}} p\,(\mathbf{k} + \mathbf{e_2})\,; \end{aligned} \tag{5.26}$$

$$\sum_{n \in S} p\,(\mathbf{k}) = 1. \tag{5.27}$$

In accordance with the PASTA theorem we find the blocking probability of v-calls ($\mathrm{PB_v}$) as follows

$$\mathrm{PB}_v := \sum_{k \in S} p\,(\mathbf{k})\, \delta\,(k_1, m)\,. \tag{5.28}$$

The average number of d-calls in the system (L_{d}) is calculated as follows

$$L_{\mathrm{d}} := \sum_{\xi=1}^{\infty} \xi p\,(\xi), \tag{5.29}$$

where $p(\xi) := \sum_{k \in S} p(\mathbf{k})\delta(k_2, \xi)$, $\xi = 1, 2, \ldots$, are the marginal distribution functions of the initial model.

Then average sojourn time in the system for d-calls (W_d) can be calculated from the well-known Little's formula.

Now apply the approximate approach to finding the above-mentioned QoS metrics of the given model. First assume that $\lambda_\text{v} >> \lambda_\text{d}$ and $\mu_\text{v} >> \mu_\text{d}$. This regime is a common one for many realistic voice/data networks.

Consider the following splitting of state space (5.24):

$$S = \bigcup_{i=0}^{\infty} S_i, \; S_i \bigcap S_j = \varnothing, \, i \neq j, \tag{5.30}$$

where $S_i := \{\mathbf{k} \in S : k_2 = i\}$.

Then from (5.25) we conclude that the stationary distributions of all split models with state space S_i are defined as the stationary distribution of the classical Erlang's model *M/M/m/0* with load ν_v Erl, i.e. $\theta_i(\nu_\text{v}, m)$, $i = \overline{0, m}$. This means that elements of the GM of a merged model $q(<i>, <j>), i, j = 0, 1, 2, \ldots$, are

$$q(<i>, <j>) = \begin{cases} \lambda_\text{d} & \text{if}\, j = i + 1 \\ \mu_\text{d} \sum_{i=o}^{m} i\theta_0(\nu_\text{v}, N - xi) & \text{if}\, j = i - 1 \\ 0 & \text{otherwise.} \end{cases} \tag{5.31}$$

Thus, the stationary distribution of a merged model is determined thus:

$$\pi(<i>) = a^i (1 - a), \quad i = 0, 1, 2, \ldots, \tag{5.32}$$

where

$$a = \frac{\nu_\text{d}}{\sum_{i=0}^{m} (N - xi)\theta_i(\nu_\text{v}, m)}.$$

Note that in determining formulae (5.32) the ergodicity condition of the system in the given mode of operation is $a < 1$. This condition has a quite probabilistic meaning since it represents the following interpretation: if the number of channels assigned to v-calls is i then the average number of available channels for d-calls is equal to the denominator of a, i.e. the obtained ergodicity condition is a common one for multi-channel infinite queuing systems with appropriate loading parameters.

In summary the following simple formulae for calculation of the desired QoS metrics can be suggested:

$$\text{PB}_\text{v} = E_\text{B}(\nu_\text{v}, m), \tag{5.33}$$

$$L_\text{d} = a/(1 - a). \tag{5.34}$$

Note that formula (5.33) might be proposed directly without any investigation since v-calls are handled in a pre-emptive manner, however, it is also exactly obtained as a result of applying the given algorithm. So this fact is indirect evidence of the correctness of the proposed approach. In addition, formula (5.34) is very similar to the one for the classical *M/M/1* queue with load a Erl, but for the given system this QoS metric is a less than appropriate one for the classical system because it occupies all available channels simultaneously.

For the finite capacity model the function PB_v will also be calculated by (5.33) but in this case a new QoS metric arises – the d-call blocking probability (PB_d). The latter and other related QoS metrics are defined as following:

$$PB_d(R) = a^R(1-a)\,/\left(1-a^{R+1}\right),$$

$$L_d(R) = \frac{1-a}{1-a^{R+1}}\sum_{k=1}^{R} ka^k,$$

$$W_d(R) = L_d(R)\,/\lambda_d(1-P_d)$$

where $L_d(R)$ and $W_d(R)$ are the mean number of d-calls in the system and mean sojourn time in the system for d-calls in the finite capacity model, respectively (in this model as mentioned above the ergodicity condition is not required), and R is the total capacity of the system for d-calls (in channels plus in the buffer).

Note that the formulae developed above assume that traffic parameters satisfy the conditions: $\lambda_v >> \lambda_d$ and $\mu_v >> \mu_d$. Appropriate formulae can also be developed for systems in which the inverse ratios are fulfilled, i.e. now we shall consider systems in which $\lambda_v << \lambda_d$ and $\mu_v << \mu_d$.

In this case the following splitting of state space (5.24) is considered:

$$S = \bigcup_{i=0}^{m} S_i,\; S_i \bigcap S_j = \varnothing, i \neq j, \tag{5.35}$$

where $S_i := \{\mathbf{k} \in S : k_1 = i\}$.

Here splitting (5.35) is based on fixing the value of the first component of a state vector. Unlike splitting (5.30) the corresponding merge function in this case determines the finite-dimensional merged model with the state space $\{<i>: i = 0, 1, \ldots, m\}$.

In this case the stationary distribution within class S_i coincides with the stationary distribution of the classical system $M/M/1/\infty$ with a load $v_i := \frac{v_d}{N-ix}$, $i = \overline{0, m}$. For the existence of a stationary mode in the split model with state space S_i the condition $v_i < 1$ must be fulfilled. As long as these conditions are fulfilled for any i we discover an ergodicity condition for the initial model under the given mode of operation: $v_d < N - mx$.

It is important to note that in the given operation mode the ergodicity condition of the model does not depend on the load of v-calls but depends on their bandwidth parameter x only. In the special case $N \bmod x = 0$ the above-mentioned existence condition for ergodicity is not satisfied for any admissible (nonnegative) values of d-call loads. These facts should be expected since unbuffered v-calls have pre-emptive-resume priorities over d-calls and the following conditions are satisfied: $\lambda_v << \lambda_d$ and $\mu_v << \mu_d$.

Also note that in the last operation mode the ergodicity condition (i.e. $v_d < N–mx$) is a stronger limiting condition than the previous one (i.e. $a < 1$). This is explained by the fact that in this case the rate of buffered d-calls is essentially higher than the rate of unbuffered v-calls.

In view of (5.25) it can be seen that by usage of the splitting (5.35) the elements of the GM of the merged model are determined as follows:

$$q(<i>,<j>) = \begin{cases} \lambda_{v}, & \text{if} j = i+1, \\ k\mu_v & \text{if} j = i-1, \\ 0 & \text{in other cases.} \end{cases} \tag{5.36}$$

From (5.36) it is concluded that the stationary distribution of the merged model in this case coincides with the stationary distribution of the classical Erlang's model *M/M/m/0* with load v_v Erl.

The QoS metric for v-calls in such a mode of operation is determined as follows:

$$\mathrm{PB_v} \approx \pi(<m>) = \theta_{\mathrm{m}}(v_{\mathrm{v}}, m) = E_{\mathrm{B}}(v_{\mathrm{v}}, m).$$

In other words, again for an evaluation of $\mathrm{PB_v}$ the formula (5.33) is used.

To find the average number of d-calls in the system in such a mode of operation we obtain the following formula:

$$L_{\mathrm{d}} \approx \sum_{j=1}^{\infty} j \sum_{i=0}^{m} \theta_i(v_{\mathrm{v}}, m)(1 - v_i) v_i^j,$$

or after some mathematical transformations we obtain the expression

$$L_{\mathrm{d}} \approx \sum_{i=0}^{m} \frac{v_i \theta_i(v_{\mathrm{v}}, m)}{1 - v_i}. \tag{5.37}$$

Further from (5.37) an average response time of d-calls in the system (W_{d}) can be found.

Formulae for calculation of QoS metrics of heterogeneous calls under such a condition can be similarly derived, also in models with limited buffers for waiting d-calls. So, for this model $\mathrm{PB_v}$ is also determined by (5.33), and the average number of d-calls in the system is determined thus

$$L_{\mathrm{d}}(R) \approx \sum_{i=1}^{R} i \sum_{j=0}^{m} \theta_j(v_{\mathrm{v}}, m)\left(1 - v_j\right) v_j^i.$$

5.3 Numerical Results

On the basis of the algorithms proposed in Sects. 5.1 and 5.2 computational programs for studying the behavior of the characteristics of models studied herein were developed. For the sake of brevity, only results of numerical experiments performed for models with a continuous band (see Sect. 5.1) are presented in Figs. 5.1, 5.2, 5.3, 5.4, and 5.5.

The following conclusions for models with a continuous band for d-calls may be drawn from an analysis of the results of numerous experiments:

- the blocking probability function of both types of calls decreases monotonically relative to the number of channels in the system (see Figs. 5.1 and 5.2), while the mean number of busy channels, on the other hand, increases monotonically as the number of channels increases (Fig. 5.3) whatever the load on the system. These results fully correspond to theoretical expectations;
- the probability function of the degradation states increases monotonically relative to the threshold parameter δ (Fig. 5.4), while the complementary function, i.e. the function that estimates the probability that d-calls are serviced with a required service rate, on the other hand, decreases monotonically relative to this argument δ. And these results possess an entirely reasonable meaning;
- as expected, the probability function of degradation of a d-call for any load and any number of channels increases monotonically relative to parameter δ (Fig. 5.5).

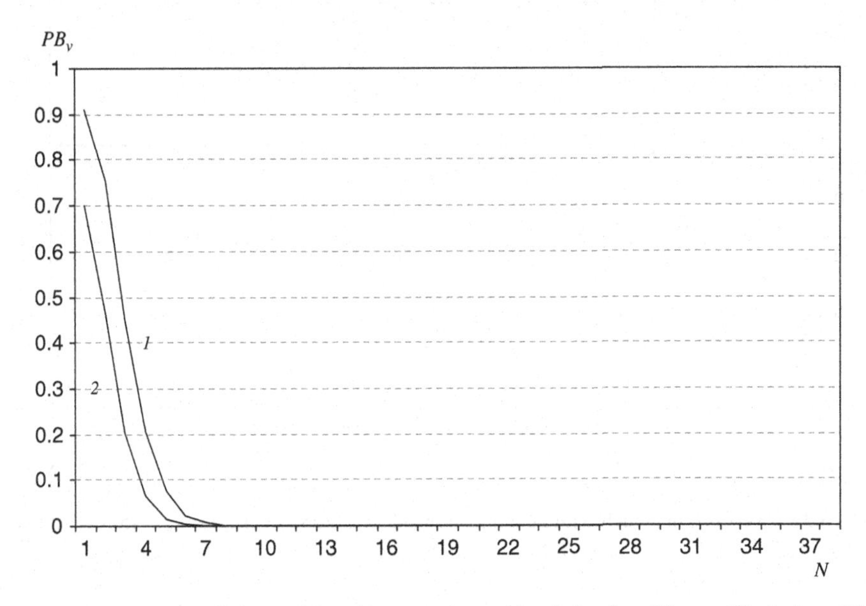

Fig. 5.1 $\mathrm{PB_v}$ versus N in models with a continuous band for $\delta = 25$, $\mu = 10$; $1 - \lambda_v = 15$, $\lambda_d = 0.01$; $2 - \lambda_v = 10$, $\lambda_d = 0.1$

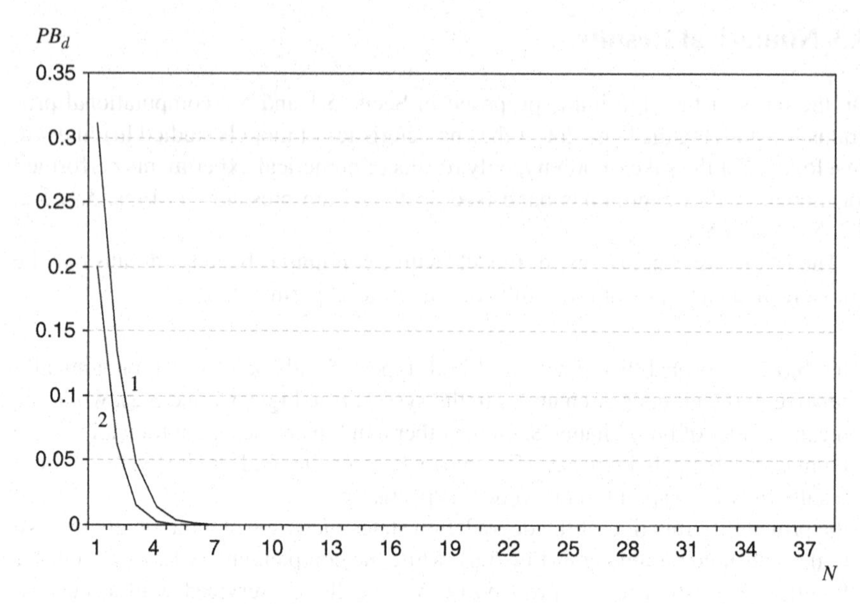

Fig. 5.2 PB_d versus N in models with a continuous band for $\delta = 25$, $\mu = 10$; $1 - \lambda_v = 15$, $\lambda_d = 0.01$; $2 - \lambda_v = 10$, $\lambda_d = 0.1$

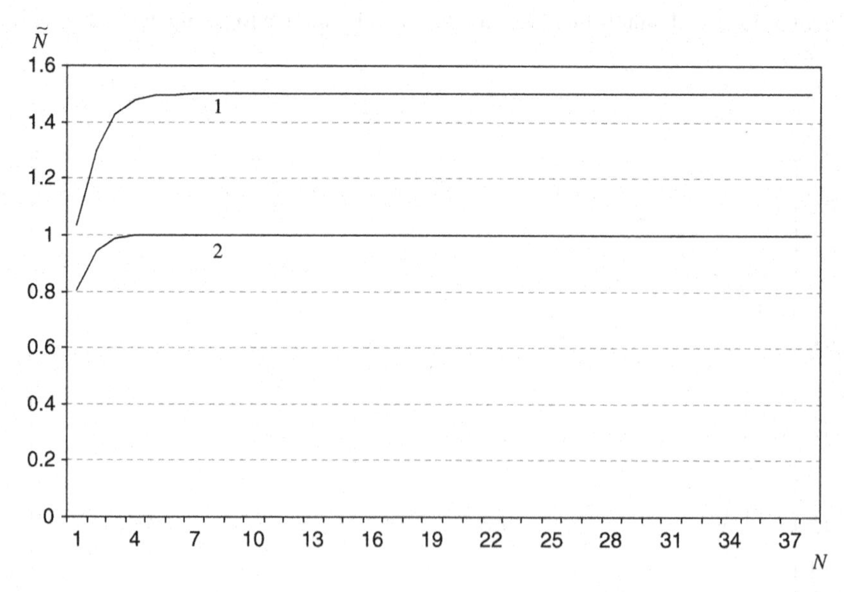

Fig. 5.3 $\tilde{N}$ versus N in models with a continuous band for $\delta = 25$, $\mu = 10$; $1 - \lambda_v = 15$, $\lambda_d = 0.01$; $2 - \lambda_v = 10$, $\lambda_d = 0.1$

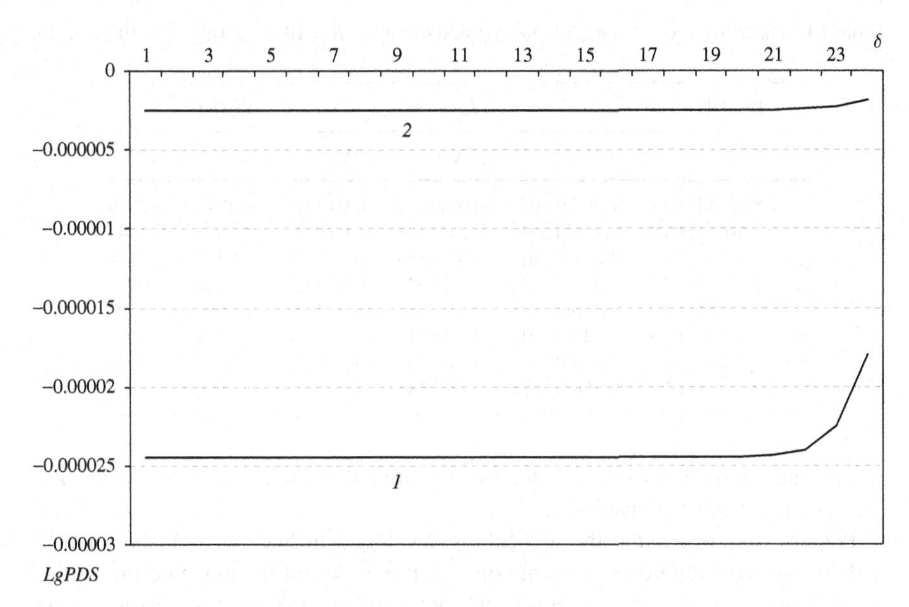

Fig. 5.4 PDS versus δ in models with a continuous band for $N = 25$, $\mu = 10$; 1 – $\lambda_v = 10$, $\lambda_d = 0.1$; 2 – $\lambda_v = 10$, $\lambda_d = 0.01$

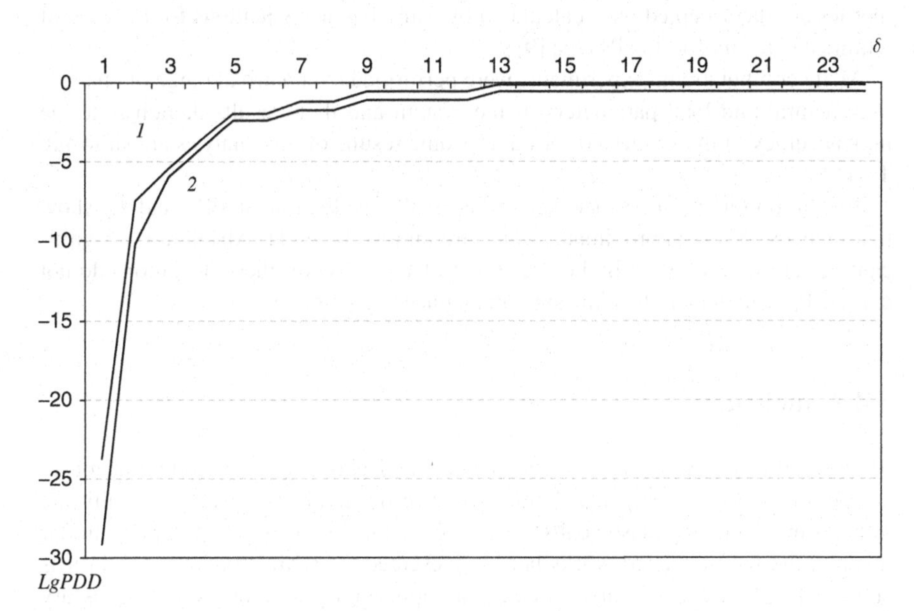

Fig. 5.5 PDD versus δ in models with a continuous band for $N = 25$, $\mu = 10$, $\lambda_v = \mu_v = \mu_d = 10$; 1 – $\lambda_d = 0.1$; 2 – $\lambda_d = 0.01$

Table 5.1 The analysis of accuracy of the proposed formulas, $R=10$, $\lambda_v=100$, $\lambda_d=10$, $\mu_v=20$, $\mu_d=1$, $x=0.4$

	$PB_d(R)$		$L_d(R)$		$W_d(R)$	
N	EV	AV	EV	AV	EV	AV
8	5.5738E-014	1.0905E-014	0.042947	0.042017	4.2947	4.2017
10	1.4021E-015	6.7269E-016	0.031823	0.031447	3.1823	3.1447
12	1.1818E-016	7.6317E-017	0.025315	0.025126	2.5315	2.5126
14	1.7155E-017	1.2788E-017	0.021029	0.020921	2.1029	2.0921
16	3.4739E-018	2.8106E-018	0.017989	0.017921	1.7989	1.7921
18	8.8540E-019	7.5422E-019	0.015719	0.015674	1.5719	1.5674
20	2.6748E-019	2.3587E-019	0.013959	0.013928	1.3959	1.3928
22	9.2145E-020	8.3268E-020	0.012554	0.012531	1.2554	1.2531

The analogous characteristics of a model with a discrete band for d-calls exhibit the same types of relationships.

For any size of system the algorithms developed in Sect. 5.2 can be realized without any computational problem since here we do not demonstrate the results of numerical experiments relating to the behavior of QoS metrics versus system parameters. But we show here the results of numerical experiments which were performed to verify the accuracy of the developed formulae. Exact values of QoS metrics are the accepted ones calculated by using balance equations for the case of a limited capacity for d-calls (see [9]).

Many computational experiments were performed over a broad range of changes in structural and load parameters of the system and their results demonstrate the high accuracy of the suggested formulae. Some results of this analysis are shown in Table 5.1.

It is important to note that the results of the evaluation of PB_d and L_d show that their exact and approximate values are almost identical. Absolutely insignificant deviations take place in the evaluation of W_d, however these deviations do not exceed 1% which is quite admissible in engineering practice.

5.4 Conclusion

Mixed multi-rate systems in which inelastic and elastic calls are simultaneously served are poorly investigated. Some works in the field of research of multi-rate systems in which only elastic calls are served are known. So, in [1], Chap. 3, similar models (the list of related works here is presented also) are studied. Therein it is supposed that each newly arrived call at an opportunity of allocation of necessary network resources uses a bandwidth of size r_{max}, i.e. the call is served with the best quality. If at this moment the network cannot provide such quality the call adapts to the current situation and is accepted into service with an admissible bandwidth r_d where $r_{min} \le r_d \le r_{max}$. If the network is in a heavily loaded condition and cannot provide a bandwidth r_{min} one or several calls which are being served with the best quality can transfer a part of their bandwidth for acceptance of the newly arrived call. If the network is loaded in such a manner that no reassignment of channels is allowed to accept the newly arrived call it is lost. After a call leaves the system, the released bandwidth can be used by other calls with the purpose of improvement of the quality of their service. It is shown that the operation of such a system can be described by means of the classical one-dimensional Erlang model.

Models of mixed multi-rate systems have been investigated in [2, 6, 7, 10–13]. In [2, 6, 7] specific models of systems in which elastic calls (packages of GPRS data) can demand 1, 2, or 3 channels are studied. Here it is assumed that on arrival of the inelastic call (voice call) one channel occupied by an elastic call can be switched to service of the newly arrived inelastic call if at this moment all channels of the system are busy. In the mentioned studies it was assumed that the released channels remain free and are not used for service of elastic calls.

In [10] the above-mentioned restrictions are removed and general models considered. Here it is supposed that elastic calls can demand an arbitrary number of channels and reassignment of a channel for service of a newly arrived inelastic call is carried out from the elastic call with the maximal bandwidth. If all elastic calls are served by one channel reassignment does not occur and the newly arrived inelastic call is lost. The released channels are used for service of elastic calls, thus the following scheme is accepted: the released channels are first of all allocated to calls with the worst degradation. These results are generalized in [11] for models with a limited queue of inelastic calls.

In [12, 13] the models are similar to the models investigated in Sect. 5.1.2, but with a limited queue of elastic calls and absolute priorities of inelastic calls are investigated therein.

Let's note that in all known works devoted to mixed models of multi-rate systems, for calculation of their QoS metrics methods based on the solution of the corresponding SBE are used. As is known such an approach is effective only for models of small dimension.

In this chapter simple algorithms for calculating the approximate values of QoS metrics of models of multi-rate systems with elastic and inelastic calls have been developed. Two types of models are distinguished: models with a continuous band and models with a discrete band for servicing elastic calls [8]. For these types of unbuffered models new types of QoS metrics are introduced to complement the existing performance measures (blocking probability of different types of calls and the mean number of busy channels), i.e. the probability of degradation states and the probability of degradation of elastic calls. Also models with either limited or unlimited queues of elastic calls have been considered. Corresponding computational programmes were developed for all these types of models and numerous computational experiments over a broad range of variation of the load and structure parameters of the models were carried out on the basis of these programmes. Through use of the formulae determined in the study, it is possible to solve important problems in relation to the selection of optimal service regimes for different types of calls. A few such problems of different types are investigated in the next chapter.

References

1. Chen H, Huang L, Kumar S, Kuo CC (2004) Radio resource management for multimedia QoS support in wireless networks. Kluwer, Boston, MA
2. Chen WY, Wu JLC, Lu LL (2003) Performance comparisons of dynamic resource allocation with/without channel de-allocation in GSM/GPRS networks. IEEE Commun Lett 7(1):10–12
3. ISO/IEC JTC1/SC29/WG11 Generic coding of moving pictures and associated audio information, ISO/IEC International Standard 13818, Nov. 1994
4. ISO/IEC JTC1/SC29/WG11 JPEG2000 Part 1 final committees draft version 1, ISO/IEC International Standard N1646, Mar. 2000
5. ISO/IEC JTC1/SC29/WG11 Overview of the MPEG-4 standard, ISO/IEC International Standard N3747, Oct. 2000
6. Lin P, Lin YB (2001) Channel allocation for GPRS. IEEE Trans Veh Technol 50(2):375–387
7. Lin P (2003) Channel allocation for GPRS with buffering mechanisms. Wirel Netw 9:431–441
8. Melikov AZ, Nagiev FN, Kaziev TS (2007) Approximate calculation of the characteristics of multi-rate systems with inelastic and elastic calls. Autom Control Comput Sci 41(4):203–214
9. Servi LD (2002) Algorithmic solutions for two-dimensional birth-death processes with application to capacity planning. Telecommun Syst 21(2–4):205–212
10. Zhang Y, Soong BH (2004) Performance evaluation of GSM/GPRS networks with channel re-allocation scheme. IEEE Commun Lett 8(5):280–282
11. Zhang Y, Soong B, Ma M (2006) A dynamic channel assignment scheme for voice/data integration in GPRS networks. Comput Commun 29(9):1163–1173
12. Zhuang W, Bensaou B, Chua KC (1999) Adaptive quality of service handoff priority scheme for wireless mobile multimedia networks. IEEE Trans Veh Technol 40(2):494–505
13. Zhuang W, Bensaou B, Chua KC (2000) Handoff priority scheme with preemptive, finite queueing and reneging in mobile multi-service networks. Telecommun Syst 15:37–51

Chapter 6
Parametric Optimization Problems in Multi-Rate Systems

Some access strategies investigated in the previous Chaps. 4 and 5 contain controllable parameters. In other words, by the choice of suitable values for these parameters, it is possible to optimize the QoS metrics of the investigated systems. Such problems we shall name problems of parametric optimization. In the present chapter similar problems for multi-rate queues (MRQ) are considered and algorithms are proposed for their solution. Realization of the results of the considered problems in real multimedia systems does not present any difficulties.

The problems considered herein for general models of MRQ with arbitrary numbers of traffic represent difficult mathematical problems. Therefore, parametric optimization problems for both Gimpelson-type models with and without buffers are investigated in detail. A similar problem for a mixed MRQ model is also investigated. The results of numerical experiments for the considered problems and their detailed analysis are given.

In this chapter we maintain all notations of the two previous chapters.

6.1 Problems for Unbuffered Gimpelson's Models

In this section problems of parametric optimization for unbuffered Gimpelson-type models are considered for three different access strategies. Thus, for the CS (Complete Sharing) access strategy the problem of the choice of the minimal rate of service at which the given restrictions on blocking probabilities of heterogeneous calls are satisfied is solved. For the TR (Trunk Reservation) access strategy various problems of the choice of optimal values of threshold parameters of narrow-band and wide-band calls are solved. Here criteria of optimality are the weighed sum of blocking probabilities of heterogeneous calls and utilization of channels. And lastly, for the SGC (Special Group of Channels) access strategy the problem of finding extreme values of a parameter that limits the number of w-calls in the common zone of channels is solved. Thus, satisfaction of restrictions on blocking probabilities of heterogeneous calls is required.

L. Ponomarenko et al., *Performance Analysis and Optimization of Multi-Traffic on Communication Networks*, DOI 10.1007/978-3-642-15458-4_6,

6.1.1 Problem of Equivalent Capacity with the CS-Strategy

When using the CS access strategy in MRQ there are only a very limited number of controllable parameters (see Sect. 4.1.1). Frequently in real systems it is very difficult to control the input traffic (i.e. depending on the current situation dynamically control the rate of traffic) though in some works such problems have been investigated. Therefore, for this access strategy problems of improvement of desirable QoS metrics based on a choice of appropriate values of service rates of calls (especially by means of changes in the value of parameter m which specifies the number of channels required for the service of w-calls) and numbers of channels are real.

The analysis of numerical experiments for QoS metrics of multi-rate queues with the CS-strategy shows their non-ordinary behavior. This especially concerns the function PB_n in all admissible areas of change of both structural and loading parameters of the model (see Sect. 4.1.4). The latter facts complicate the formulation and solution of optimization problems for the MRQ model on using the given access strategy.

At the same time, for the given system, from both theoretical and practical points of view, the solution of the problem of equivalent capacity is interesting. This problem is formulated as follows: for the given number of channels and for known rates of heterogeneous calls it is required to find such minimal values of service rates that blocking probabilities do not exceed the given limits.

First we shall assume, that speed of the channel is a continuous quantity and let equation $\mu_w = m\mu_n$ be valid. The last assumption reflects a real situation as w-calls are served simultaneously by m channels and it is natural to assume that their service rate in m time surpasses the service rate of n-calls.

Now we shall consider the mathematical statement of this problem. This is written as follows:

$$\mu_n \to \min \tag{6.1}$$

$$\text{s.t.} \quad PB_w(CS) \le \varepsilon_w, \tag{6.2}$$

$$PB_n(CS) \le \varepsilon_n, \tag{6.3}$$

where ε_n and ε_w are the given upper limits for blocking probabilities.

Let's denote the optimal solution of problem (6.1), (6.2), and (6.3) through μ^*. For the solution of this problem in view of the monotonic properties of functions on the left sides of constraints (6.2) and (6.3) it is possible to use the dichotomy algorithm. For this purpose it is required to define preliminary limits of an interval $[\underline{\mu}, \bar{\mu}]$ containing μ^*. In particular, these limits can be found as the solutions of parallel problems of equivalent capacity for two auxiliary minor and major (with respect to the initial model) systems, i.e. $\underline{\mu}(\bar{\mu})$ is the solution of the mentioned problem for minor (major) queuing system $M|M|N|0$ $(M|M|[N/m]|0)$ with input rate $\lambda_n(\lambda_n+\lambda_w)$ and with a required threshold for blocking probability. This threshold is given by $\varepsilon = \min\{\varepsilon_n, \varepsilon_w\}$. The problem of equivalent capacity for queuing systems

Table 6.1 Results of the problem (6.1), (6.2), and (6.3)

N	60	60	60	50	50	50	40	40	40	30
m	6	6	12	6	12	12	12	12	6	6
λ_n	50	40	30	30	30	50	45	60	60	50
λ_w	50	30	40	40	30	60	40	40	60	40
ε_n	10^{-4}	10^{-5}	10^{-5}	10^{-6}	10^{-6}	10^{-7}	10^{-5}	10^{-4}	10^{-5}	10^{-5}
ε_w	10^{-5}	10^{-5}	10^{-5}	10^{-6}	10^{-6}	10^{-7}	10^{-4}	10^{-4}	10^{-6}	10^{-6}
μ^*	7.69	5.04	23.86	11.91	46.52	141.95	40.83	44.07	34.46	77.93

with homogenous calls is easily solved, and consequently the solution of problems of finding $\underline{\mu}$ and $\bar{\mu}$ do not represent fundamental difficulties.

Some results of the solution of problem (6.1), (6.2), and (6.3) using the specified approach are shown in Table 6.1, thus the error of finding μ^* does not exceed 0.1 %.

The analysis of the results of the solution of problem (6.1), (6.2), and (6.3) shows that with growth of traffic loads and width of w-calls the optimal value of μ^* grows, and with the increase of ε_n and/or ε_w it decreases.

In the case where the speeds of the channel can only accept a finite number of discrete values, problem (6.1), (6.2), and (6.3) becomes even simpler, and in such cases the optimal solution of the given problem (if it exists) is defined precisely.

In summary we shall note, that the solution of a similar problem for a system in which the CSE-strategy is used, becomes much simpler, as with use of this strategy the blocking probabilities of heterogeneous calls equal each other. Therefore, the similar problem of optimization for the CSE-strategy is not considered here.

6.1.2 Problems of Finding the Optimal CAC Parameters with the TR-Strategy

First consider a special Gimpelson-type model in which $\mathbf{b} = \left(\underbrace{1,\ldots,1}_{k_1},\underbrace{N,\ldots,N}_{k_2}\right)$. It is clear that for this model, a feasible value of r_w is zero and here we assume that $0<r_n\leq N-1$, since if $r_n = 0$ we get the already obtained results for the CS-strategy.

By using the algorithm proposed in Sect. 4.1.3 we obtain the following formulae for calculation of the QoS metrics PB_n and PB_w:

$$PB_n(r_n) = \hat{\nu}_n\pi(0),\ PB_w(r_n) = 1 - \pi(0), \tag{6.4}$$

where

$$\hat{\nu}_n := \sum_{i=1}^{k_1} \nu_i,\ \hat{\nu}_2 := \hat{\nu}_3 := \ldots = \hat{\nu}_{N-1} := 0,\ \hat{\nu}_w := \sum_{i=k_1+1}^{k_1+k_2} \nu_i,\ \pi(0) = \left(\hat{\nu}_w + \sum_{i=0}^{N-r_n} \frac{\hat{\nu}_n^i}{i!}\right)^{-1}.$$

For large N when $r_n << N$ we may use an approximate simple formula to calculate blocking probabilities since in this case $\pi(0) \approx (\exp(\hat{\nu}_n) + \hat{\nu}_w)$. From formulae (6.4) we conclude that:

(i) if the values of $\hat{\nu}_n$, $\hat{\nu}_w$, and r_n are fixed, then when N increases, $PB_w(r_n)$ also increases and $PB_n(r_n)$ decreases;
(ii) if the values of $\hat{\nu}_n$, $\hat{\nu}_w$, and N are fixed, then when r_n increases, $PB_n(r_n)$ also increases and $PB_w(r_n)$ decreases;
(iii) $PB_w(r_n) > PB_n(r_n)$ for any $\hat{\nu}_n$, $\hat{\nu}_w$, r_n, and N, since this inequality is equivalent to $\sum_{i=0}^{N-r_n} \frac{\hat{\nu}_n^i}{i!} > 1$ that is valid for any possible values of the mentioned parameters;
(iv) $\min_{r_n} \{PB_w(r_n) - PB_n(r_n)\} = PB_w(N-1) - PB_n(N-1) = \hat{\nu}_n / \left(1 + \hat{\nu}_n + \hat{\nu}_w\right)$.

Now consider the problem of finding the optimal value of $r_n{}^*$ for minimization of the weighted sum of blocking probabilities (WPB):

$$\mathrm{WPB}\,(r_n) := \frac{\hat{\nu}_w}{\hat{\nu}_n + \hat{\nu}_w} \cdot PB_w(r_n) + \frac{\hat{\nu}_n}{\hat{\nu}_n + \hat{\nu}_w} \cdot PB_n(r_n) \underset{r_n}{\longrightarrow} \min . \tag{6.5}$$

By using formulae (6.4) it can easily be shown that problem (6.5) is equivalent to:

$$\sum_{i=0}^{N-r_n} \frac{\hat{\nu}_n^i}{i!} + \hat{\nu}_w \underset{r_n}{\longrightarrow} \max .$$

It is clear that the last problem has the optimal solution $r_n^* = 1$ for any possible values of $\hat{\nu}_n$, $\hat{\nu}_w$, and N.

In an analogous way it can easily be shown that the optimal value of r_n^* for maximization of channel utilization is 1 for any possible values of $\hat{\nu}_n$, $\hat{\nu}_w$, and N. Therefore, we may conclude that for the given special type of Gimpelson model the optimal value of the TR parameter for narrow-band traffic is 1 in both the unconstrained minimization of $\mathrm{WPB}(r_n)$ and maximization of $\tilde{N}(\mathrm{TR})$ problems.

The latter results and results (i)–(iv) mentioned above motivate us to organize fair servicing of heterogeneous calls in the order ε, $0<\varepsilon<1$. For this, we need to find at a fixed value of N, the minimal value of r_n such that

$$PB_w(r_n) - PB_n(r_n) \le \varepsilon, \tag{6.6}$$

where $\frac{\hat{\nu}_n}{1+\hat{\nu}_n+\hat{\nu}_w} < \varepsilon < 1$ [see (iv)].

The inequality (6.6) easily leads us to the solution with respect to x, $x = N - r_n$, of the following problem:

$$\sum_{i=0}^{x} \frac{\hat{\nu}_n^i}{i!} \le \frac{1+\hat{\nu}_w}{1-\varepsilon} - \hat{\nu}_w. \tag{6.7}$$

In solving the latter problem, it is useful to consider the following fact: the left-hand side is a monotonically increasing function with respect to x. Thus, for finding the maximal value of x (which corresponds to the minimal value of r_n), that satisfies condition (6.7), we may use, for example, the binary search method. The general scheme of the method is as follows: we take the midpoint [N/2] of the interval (1, N–1). If at this point, condition (6.7) is satisfied, then we consider a new interval ([N/2], N–1); otherwise we consider the interval (1, [N/2]). The condition that needs to be checked for to end the procedure is finding the interval unit length for which the left endpoint satisfies condition (6.7), while the right endpoint does not satisfy condition (6.7). Then, this left endpoint is the desirable value of x^*, that is $r_n^* = N - x^*$. The algorithm is finite since as mentioned above, the left-hand side of (6.7) increases monotonically with respect to x. Following from results (i)–(iv) mentioned above when ε increases, r_n^* will decrease.

Now consider the general Gimpelson-type models. As before (see Sect. 4.1.4) for distinctness and without loss of generality, we assume that $b_n = 1$ and $b_w = m$, $1<m<N$. Unlike the special type of Gimpelson model in this case it is possible to reach absolute fair servicing of heterogeneous calls in the sense of equality of their blocking probabilities. Indeed, from the results of Sect. 4.1.3 we conclude that for any possible values of ν_n, ν_w, and N the following relationship is valid:

$$PB_n(r_n, r_w) = PB_w(r_n, r_w), \text{ if } r_n - r_w = m - 1. \tag{6.8}$$

The last fact has a simple intuitive explanation. So if condition $r_n - r_w = m - 1$ is valid then both types of calls have the same width, i.e. in this case both types of calls are accepted if the number of free channels is $m+r_w = r_n+1$. Moreover, from the results of Sect. 4.1.3 we conclude that for any possible values of ν_n, ν_w, and N the following relationships are valid:

$$PB_n(r_n, r_w) < PB_w(r_n, r_w) \text{ if } r_n < r_w + m - 1; \tag{6.9}$$

$$PB_n(r_n, r_w) > PB_w(r_n, r_w) \text{ if } r_n > r_w + m - 1. \tag{6.10}$$

It is clear that for the case when traffic is symmetric in the sense of bandwidth requirement ($b_n = b_w$), the problems of minimization of the weighted sum of blocking probabilities and maximization of channel utilization are equivalent. But for cases where networks have asymmetric traffic in the mentioned sense ($b_n \neq b_w$), the optimal strategy for solution indicated for the above two optimization problems may in general differ. If we wish to maximize channel utilization, we should give w-calls an advantage which will in turn lead to worsening blocking probabilities for n-calls and vice versa.

In order to present the last comments numerically, first we consider the unconstrained problem of finding the values of r_n and r_w at different ν_n and ν_w values, that minimizes the WPB(r_n,r_w) for the general Gimpelson-type models. In other words, here our objective is:

$$WPB\,(r_n, r_w) := \frac{\nu_w}{\nu_n + \nu_w} \cdot PB_w\,(r_n, r_w) + \frac{\nu_n}{\nu_n + \nu_w} \cdot PB_n\,(r_n, r_w) \min. \tag{6.11}$$

For the solution of this problem and for later work the following useful facts are to be used: PB_n (PB_w) increases monotonically with respect to $r_n(r_w)$ at a fixed value of $r_w(r_n)$, and when $r_w(r_n)$ increases the value of $PB_n(PB_w)$ decreases while N is fixed. In other words, the following boundaries for PB_n and PB_w can be proposed:

$$PB_n\,(0, N-m) \le PB_n\,(r_n, r_w) \le PB_n\,(N-1, 0)\,; \tag{6.12}$$

$$PB_w\,(N-1, 0) \le PB_w\,(r_n, r_w) \le PB_w\,(0, N-m)\,. \tag{6.13}$$

Therefore, for solution of problem (6.11), in particular, the following approach may be used: fix the value of either one of the TR parameters r_n or r_w consecutively, and use any one-dimensional search method to find the value of the other parameter. The existence of the solution follows from the fact that the set of possible values of r_n and r_w is finite.

The results of problem (6.11) for the sample model with $N = 30$, $b_n = 1$, $b_w = 6$ are given in Table 6.2, which shows the different regions for optimal solutions with respect to (6.11).

Now consider the following constrained problem:

$$\tilde{N}\,(r_n, r_w) \underset{r_n, r_w}{\longrightarrow} \max \tag{6.14}$$

$$\text{s.t.} \quad WPB\,(r_n, r_w) \le \varepsilon, \tag{6.15}$$

where ε is a given possible upper limit for WPB.

Numerical experiments for this problem were also carried out for the sample model with $N = 30$, $b_n = 1$, $b_w = 6$ (see Table 6.3). For a small value of $\varepsilon(\varepsilon \le 0.2)$, only for a few values of ν_w is it possible to satisfy constraint (6.15) and we obtain

Table 6.2 Results of the problem (6.11) for the sample model with $N = 30$, $b_n = 1$, $b_w = 6$ o – $r_n^* = r_w^* = 1$, x – $r_n^* = 1$, $r_w^* = 2$, Δ – $r_n^* = 1$, $r_w^* = 3$

ν_w \ ν_n	1	2	3	4	5	6	7	8	9	10
10	o	x	x	x	x	x	Δ	Δ	Δ	Δ
9	o	x	x	x	x	x	Δ	Δ	Δ	Δ
8	o	x	x	x	x	x	Δ	Δ	Δ	Δ
7	o	x	x	x	x	x	Δ	Δ	Δ	Δ
6	o	x	x	x	x	x	x	Δ	Δ	Δ
5	o	x	x	x	x	x	x	Δ	Δ	Δ
4	o	o	x	x	x	x	x	Δ	Δ	Δ
3	o	o	x	x	x	x	x	x	Δ	Δ
2	o	o	o	x	x	x	x	x	x	Δ
1	o	o	o	x	x	x	x	x	x	x

Table 6.3 Results of optimal strategies of problems (6.14) and (6.15) for the sample model: (a) $-\varepsilon = 0.4$, (b) $-\varepsilon = 0.5$

(a)

v_w \ v_n	1	2	3	4	5	6	7	8	9	10
10	Ø	Ø	Ø	Ø	Ø	Ø	Ø	Ø	Ø	Δ
9	Ø	Ø	Ø	Ø	Ø	Ø	Ø	□	□	□
8	Ø	Ø	Ø	Ø	□	□	□	o	o	o
7	Ø	Ø	o	o	o	o	o	o	o	o
6	Ø	o	o	o	o	o	o	o	o	o
5	o	o	o	o	o	o	o	o	o	o
4	o	o	o	o	o	o	o	o	o	o
3	o	o	o	o	o	o	o	o	o	o
2	o	o	o	o	o	o	o	o	o	o
1	o	o	o	o	o	o	o	o	o	o

(b)

v_w \ v_n	1	2	3	4	5	6	7	8	9	10
10	Ø	Ø	Ø	□	o	o	o	o	o	o
9	Ø	Ø	o	o	o	o	o	o	o	o
8	Ø	o	o	o	o	o	o	o	o	o
7	o	o	o	o	o	o	o	o	o	o
6	o	o	o	o	o	o	o	o	o	o
5	o	o	o	o	o	o	o	o	o	o
4	o	o	o	o	o	o	o	o	o	o
3	o	o	o	o	o	o	o	o	o	o
2	o	o	o	o	o	o	o	o	o	o
1	o	o	o	o	o	o	o	o	o	o

the same optimal strategy as in the appropriate unconstrained problem since in these cases the admissible set of problem (6.14) and (6.15) becomes very small. Some other strategies obtained from problem (6.14) and (6.15) are shown in Table 6.3, where we maintain the notations of Table 6.2. Note that in the above problem, the results of problem (6.11) was very useful in determining the admissible set.

As is indicated by these tables, optimal strategies vary with changes in the ε value. At low values of ε no-solution cases increase as the requirement is too limiting to be satisfied. When ε increases the optimal strategy area $r_{\text{n}}^{*}=r_{\text{w}}^{*}=1$ moves to the top and note that eventually when ε becomes large enough, the optimal solution of the constrained problem (6.14) and (6.15) becomes the same as the solution of the appropriate unconstrained problem as expected.

An analogous approach can be used to consider the duality of the (6.14) and (6.15) optimization problem, that is:

$$\mathrm{WPB}\,(r_{\mathrm{n}}, r_{\mathrm{w}}) \underset{r_n, r_w}{\longrightarrow} \min$$

$$\text{s.t.} \quad \tilde{N}\,(r_{\mathrm{n}}, r_{\mathrm{w}}) \geq \underline{N},$$

where $\underline{N}$ is a given possible lower limit for $\tilde{N}\,(r_{\mathrm{n}}, r_{\mathrm{w}})$.

Usually QoS requirements for heterogeneous traffic are defined by giving upper bounds for blocking probabilities. One possible way of stating this problem can be formulated as follows: find the extreme (minimal or maximal) values of r_{n} and r_{w} such that

$$\mathrm{PB}_{\mathrm{n}}(r_{\mathrm{n}}, r_{\mathrm{w}}) \leq \varepsilon_{\mathrm{n}} \tag{6.16}$$

$$\mathrm{PB}_{\mathrm{w}}(r_{\mathrm{n}}, r_{\mathrm{m}}) \leq \varepsilon_{\mathrm{w}}. \tag{6.17}$$

It is clear that ε_{n} and ε_{w} should correspond to (6.12) and (6.13). To make a concrete statement we shall consider the problem of definition of the minimal values of the specified parameters. The general scheme of the proposed procedure is as follows. First, in consecutive order starting from the maximal possible value, for example, of r_{w} for each of its fixed values, by using any one-dimensional search method for the other parameter (i.e. r_{n}), we obtain the admissible set of solutions for condition (6.16). Second, we choose from the above-defined set such pairs (r_{n}, r_{w}) at fixed values of r_{w} with maximal values of r_{n}. If condition (6.17) does not hold for any such pairs then the problem has no solution. Otherwise, that is, if condition (6.17) holds for at least one pair (r_{n}, r_{w}) from the admissible set, then we choose among these pairs the pair with the minimal value of r_{w}. Third, for this pair, we use a one-dimensional search method for parameter r_{n}, to find the solution of problem (6.17) and therefore also (6.16) and (6.17).

Note that the proposed procedure can be modified for solution of the problem of finding the maximal values of r_{n} and r_{w}. Thus, by combining the solutions of these problems we can find all admissible values of r_{n} and r_{w} at which constraints (6.16) and (6.17) are carried out.

Numerous numerical experiments using the proposed algorithms for the sample model in problem (6.11) were carried out, where ε_{n} and ε_{w} are chosen in accordance with (6.12)and (6.13) as follows: $\varepsilon_{\mathrm{n}} = k_{\mathrm{n}}\mathrm{PB}_{\mathrm{n}}(0{,}24){+}(1{-}k_{\mathrm{n}})\mathrm{PB}_{\mathrm{n}}(29{,}0)$, $\varepsilon_{\mathrm{w}} = k_{\mathrm{w}}\mathrm{PB}_{\mathrm{w}}(29{,}0) + (1{-}k_{\mathrm{w}})\mathrm{PB}_{\mathrm{w}}(0{,}24)$, $0{\leq}k_x{\leq}1$, $x{\in}\{n,w\}$.

Note that in the case $k_{\mathrm{n}} = k_{\mathrm{w}} = 0.8$ we obtain the fixed optimal strategy $r_{\mathrm{n}}^* = r_{\mathrm{w}}^* = 1$ for all ν_{n} and ν_{w} where $1{\leq}\nu_{\mathrm{n}}{\leq}7$ and $1{\leq}\nu_{\mathrm{w}}{\leq}10$. For the values of $\nu_{\mathrm{n}}{\geq}8$, a mixture of strategies is obtained. But when the requirements of w-calls are increased (that is the value of k_{w} is closer to one), for many regions of ν_{n} and ν_{w} problem (6.16) and (6.17) has no solution. Some of the results of the given problem are presented in Table 6.4, where the new notations in some of the table entries indicate the optimal (minimal) values of r_{n} and r_{w}.

Table 6.4 Results of problem (6.16) and (6.17) for the sample model with $N=30$, $b_n=1$, $b_w=6$ where $k_n=k_w=0.8$

v_w \ v_n	1	2	3	4	5	6	7	8	9	10
10	o	o	o	o	o	o	o	ø	ø	ø
9	o	o	o	o	o	o	o	ø	ø	ø
8	o	o	o	o	o	o	o	ø	ø	ø
7	o	o	o	o	o	o	o	ø	ø	ø
6	o	o	o	o	o	o	o	ø	ø	ø
5	o	o	o	o	o	o	o	(2,1)	ø	ø
4	o	o	o	o	o	o	o	(2,1)	ø	ø
3	o	o	o	o	o	o	o	(2,1)	(3,1)	ø
2	o	o	o	o	o	o	o	o	(3,1)	(4,1)
1	o	o	o	o	o	o	o	o	o	o

Now we shall consider a problem of improvement of QoS metrics for a model in which only one traffic component has the controllable reservation parameter. In this case the problem described above becomes even simpler.

Let for distinctness r_w be the controllable parameter, and n-calls have the fixed parameter of reservation (i.e. r_n is fixed). Here it is required to find extreme (minimal and maximal) values of parameter r_w so that the constraints on blocking probabilities of calls of each type were satisfied, that is:

$$\overline{r}_w - \underline{r}_w \to \max \tag{6.18}$$

$$\text{s.t.} \quad PB_n(r_w) \le \varepsilon_n, \tag{6.19}$$

$$PB_w(r_w) \le \varepsilon_w, \; 0 \le \underline{r}_w \le \overline{r}_w \le N - m. \tag{6.20}$$

For solution of the given problem there are the following useful limits of change for blocking probabilities of heterogeneous calls which are special cases of (6.12) and (6.13):

$$PB_n(N-m) \le PB_n(r_w) \le PB_n(0),\; PB_w(0) \le PB_w(r_w) \le PB_w(N-m),$$
$$\forall r_w \in [0, N-m].$$

In view of the latter relationships it is possible to offer the following algorithm for solution of the above-formulated problem.

Step 1. If $PB_n(N\text{–}m) > \varepsilon_n$ and/or $PB_w(0) > \varepsilon_w$ the problem (6.18), (6.19), and (6.20) has no solution.

Step 2. In parallel solve the following problems:

$$r_w^* := \arg\min_{r_w} \{PB_n(r_w) \le \varepsilon_n\}, \; r_w^{\bullet\bullet} := \arg\max_{r_w} \{PB_w(r_w) \le \varepsilon_w\}.$$

Table 6.5 Results of problem (6.18), (6.19), and (6.20) for the sample model with $N = 50, m = 5,\ r_{\mathrm{n}} = 0$

ν_{n}	ν_{w}	ε_{n}	ε_{w}	$[r_{\mathrm{w}}^{*}, r_{\mathrm{w}}^{**}]$
5	2	10^{-3}	10^{-1}	[1,22]
5	10	10^{-3}	10^{-1}	∅
5	5	10^{-2}	10^{-1}	[1,5]
5	5	$2{\cdot}10^{-3}$	10^{-1}	[1,5]
5	5	10^{-3}	$2{\cdot}10^{-1}$	[1,13]
5	15	10^{-4}	$7{\cdot}10^{-4}$	[4,18]
15	15	10^{-4}	$7{\cdot}10^{-1}$	[7,9]
15	10	10^{-1}	$5{\cdot}10^{-1}$	[1,4]
15	5	10^{-1}	$5{\cdot}10^{-1}$	[1,17]
15	5	10^{-3}	$3{\cdot}10^{-1}$	[3,8]
10	10	10^{-3}	$3{\cdot}10^{-1}$	∅
10	10	10^{-1}	$5{\cdot}10^{-1}$	[1,9]
10	5	10^{-1}	$5{\cdot}10^{-1}$	[1,22]
10	5	10^{-4}	$5{\cdot}10^{-1}$	[4,22]
4	5	10^{-4}	10^{-1}	[2,6]
4	5	10^{-6}	10^{-1}	[4,6]

Step 3. If $r_{\mathrm{w}}^{*} > r_{\mathrm{w}}^{**}$ the given problem has no solution. Otherwise the solution of the problem (6.18), (6.19), and (6.20) is $\underline{r}_{\mathrm{w}} := r_{\mathrm{w}}^{*}, \overline{r}_{\mathrm{w}} := r_{\mathrm{w}}^{**}$.

It is also possible to solve problems of channel maximization with use of the results of solution of problem (6.18), (6.19), and (6.20). The solution of the given problem does not represent any difficulties as the function is decreasing with respect to r_{w}, i.e. the maximal value of the given function is reached at a point r_{w}^{*}. Some results of the solution of the considered problem (6.18), (6.19), and (6.20) are shown in Table 6.5.

The analysis of results of the solution of problem (6.18), (6.19), and (6.20) allows one to conclude that with growth in the load of traffic entering the system the optimal interval decreases, and with an increase of ε_{n} and/or ε_{w} the optimal interval increases. And the optimal solution of the problem is not supersensitive concerning a change in parameter ε_{n}.

6.1.3 Problems of Finding the Optimal CAC Parameters with the SGC-Strategy

When using the given access strategy there are two controlled parameters (i.e. dimension of the special zone of channels for servicing w-calls and a parameter that restricts the number of w-calls in the common zone of channels), consequently, by selecting the values of these parameters it is possible to achieve a desired level of blocking of heterogeneous calls (see Sect. 4.2.1). To concretize the discussion here we will consider only a single optimization problem for the model, where a parameter that restricts the maximum number of w-calls in the common zone (i.e. $\overline{k}_w$) is the controllable one. In other words, it is supposed that both the total number of channels (N) and the dimension of the individual zone for servicing w-calls (A) are fixed. In addition, it is assumed that the traffic loads are known.

To emphasize the dependence of the blocking probabilities on the parameter $\bar{k}_{\mathrm{w}}$ as well as to facilitate further presentation, these probabilities will be denoted $\mathrm{PB}_{\mathrm{n}}\left(\bar{k}_{\mathrm{w}}\right)$ and $\mathrm{PB}_{\mathrm{w}}\left(\bar{k}_{\mathrm{w}}\right)$. Suppose that the performance of a system is estimated in terms of the blocking probability of different types of calls and that constraints on these QoS metrics are specified:

$$\mathrm{PB}_{\mathrm{n}}\left(\bar{k}_{\mathrm{w}}\right) \leq \varepsilon_{\mathrm{n}}, \tag{6.21}$$

$$\mathrm{PB}_{\mathrm{w}}\left(\bar{k}_{\mathrm{w}}\right) \leq \varepsilon_{\mathrm{w}}, \tag{6.22}$$

where as before ε_{n} and ε_{w} are specified quantities.

Then the parametric optimization problem is specified in the following way. For fixed N, m, A, v_{n}, and v_{w}, we wish to find a range of variation of the values of $\bar{k}_{\mathrm{w}}$ of maximal length $[\bar{k}_{\mathrm{w}}^{*}, \bar{k}_{\mathrm{w}}^{**}]$ within which conditions (6.21) and (6.22) are satisfied.

The properties of monotonicity of both functions $\mathrm{PB}_{\mathrm{n}}\left(\bar{k}_{\mathrm{w}}\right)$ and $\mathrm{PB}_{\mathrm{w}}\left(\bar{k}_{\mathrm{w}}\right)$ are used in developing an algorithm for solving the problem. Thus, the following relations hold:

$$\mathrm{PB}_{\mathrm{n}}(1) \leq \mathrm{PB}_{\mathrm{n}}\left(\bar{k}_{\mathrm{w}}\right) \leq \mathrm{PB}_{\mathrm{n}}\left(\left[\frac{N}{m} - A\right]\right), \tag{6.23}$$

$$\mathrm{PB}_{\mathrm{w}}\left(\left[\frac{N}{m} - A\right]\right) \leq \mathrm{PB}_{\mathrm{w}}\left(\bar{k}_{\mathrm{w}}\right) \leq \mathrm{PB}_{\mathrm{w}}(1). \tag{6.24}$$

Then, in view of (6.23) and (6.24), the following algorithm for solving the problem may be proposed.

Step 1. If $\mathrm{PB}_{\mathrm{n}}(1)>\varepsilon_{\mathrm{n}}$ and/or $\mathrm{PB}_{\mathrm{w}}([(N/m)–A])>\varepsilon_{\mathrm{w}}$ the problem does not have a solution.

Step 2. The following problems are solved in parallel:

$$k_{\mathrm{n}}^{*} := \arg\max\left\{\mathrm{PB}_{\mathrm{n}}\left(\bar{k}_{\mathrm{w}}\right) \leq \varepsilon_{\mathrm{n}}\right\},\ k_{\mathrm{w}}^{*} := \arg\min\left\{\mathrm{PB}_{\mathrm{w}}\left(\bar{k}_{\mathrm{w}} \leq \varepsilon_{\mathrm{w}}\right)\right\}.$$

Step 3. If $k_{\mathrm{n}}^{*}<k_{\mathrm{w}}^{*}$ the problem does not have a solution, otherwise $\bar{k}_{\mathrm{w}}^{*} := k_{\mathrm{w}}^{*}$, $\bar{k}_{\mathrm{w}}^{**} := k_{\mathrm{n}}^{*}$ will be a solution of the problem.

Note that the method of dichotomy, in particular, may be applied to solve the two problems in step 2 of the above-described algorithm. Some of the results of the computational experiments that were carried out are shown in Table 6.6, where the symbol Ø as before denotes that the problem does not have a solution. A number of conclusions follow from an analysis of the results:

- as the value of A increases the optimal interval moves to the left;
- as both ε_{n} and ε_{w} decrease, the length of the optimal interval shrinks and below certain values of ε_{n} and ε_{w} that are specified by means of (6.23) and (6.24) the problem does not, in general, have any solution;

Table 6.6 Results of problem (6.21) and (6.22)

ν_n	10	10	10	10	20	25	15	10	10	10	10
ν_w	10	10	10	10	10	10	30	10	10	10	10
A	4	7	7	7	5	5	5	5	5	10	10
ε_n	10^{-2}	10^{-2}	10^{-1}	10^{-1}	10^{-3}	10^{-3}	10^{-3}	10^{-1}	10^{-2}	10^{-1}	10^{-1}
ε_w	10^{-3}	10^{-3}	10^{-3}	10^{-4}	10^{-2}	10^{-2}	10^{-2}	10^{-2}	10^{-1}	10^{-1}	10^{-2}
$[\bar{k}_w^*, \bar{k}_w^{**}]$	[17,22]	[14,19]	[14,19]	Ø	[13,16]	Ø	Ø	[13,21]	[3,16]	[3,16]	[8,16]

– as the loads increase, the length of the optimal interval shrinks and above certain critical values of the loads, the problem does not have a solution.

Note that the problem of selecting an optimal value of the parameter A may be formulated and solved in a similar fashion.

6.2 Problems for Buffered Gimpelson's Models

Here problems of parametric optimization for models of multi-rate systems with buffers for w-calls are considered. In these models there are two controllable parameters – the threshold parameter (G) and size of buffer for w-calls (R). Remember, that in such models the newly arrived n-call is accepted if at this epoch the number of free channels is more than $mG, 1 \leq G \leq \overline{G}$ where $\overline{G}$ is defined much as (4.40).

Hence, as a result of a choice of their appropriate values, in some cases it is possible to reach a desirable level of blocking probabilities of heterogeneous calls. Thus, in view of the monotony of blocking probabilities on both parameters G and R, to concretize the statement we will consider here only one problem of parametric optimization of this model in which it is proposed threshold parameter be chosen as the controllable parameter. In other words, it is supposed, that the total number of channels and the size of the buffer for w-calls are fixed, and also traffic loadings are considered known. To emphasize the dependence of blocking probabilities of narrow-band and wide-band calls on the threshold parameter G, here the mentioned probabilities are denoted $PB_n(G)$ and $PB_w(G)$, respectively.

Let the overall performance of the system be estimated by the level of blocking probabilities of heterogeneous calls and restrictions on the specified QoS metrics are thus:

$$PB_n(G) \leq \varepsilon_n, \tag{6.25}$$

$$PB_w(G) \leq \varepsilon_w, \tag{6.26}$$

where ε_n and ε_w are the upper limits for blocking probabilities.

Then the problem of choice of optimal values of threshold parameter G is formulated as follows. At fixed values of N, m, ν_n, and ν_w, it is required to find such interval of change of values G (the maximal length) within which conditions (6.25) and (6.26) are satisfied. The left and right ends of the required interval we shall denote G_* and G^*, respectively.

For the solution of problem (6.25) and (6.26), in view of the properties of monotonicity of both functions $PB_n(G)$ and $PB_w(G)$ the dichotomy method can be applied. Some of the results of numerical experiments for the problem (6.25) and (6.26) are shown in Tables 6.7 and 6.8. In all experiments it is accepted, that $\lambda_n = 0.7$, $\lambda_w = 20$.

The problem, whose results are shown in Table 6.7, is characterized by $PB_n(1) > PB_w(1)$. Thus, it is necessary to remind ourselves, that function $PB_n(G)$ is increasing, and function $PB_w(G)$ on the contrary is a decreasing function of argument G (see Sect. 4.2.2). The performed experiments also characterized that

Table 6.7 Results of problem (6.25) and (6.26) for the sample model with $N = 22$, $m = 2$

μ_n	μ_w	ε_n	ε_w	$[G_*, G^*]$
1	9	10^{-1}	10^{-6}	[1,4]
1	9	10^{-2}	10^{-6}	[1,3]
1	9	10^{-1}	$8.15 \cdot 10^{-7}$	[2,4]
1	9	10^{-2}	$8.15 \cdot 10^{-7}$	[2,3]
2	3	10^{-1}	10^{-2}	$\varnothing$
2	7	10^{-1}	10^{-2}	[1,5]
2	7	10^{-1}	10^{-5}	[1,3]
2	9	10^{-1}	10^{-3}	[1,6]
2	9	10^{-4}	10^{-6}	$\varnothing$
2	9	10^{-1}	$2.99 \cdot 10^{-7}$	$\varnothing$
2	9	10^{-1}	$8.15 \cdot 10^{-7}$	[1,6]
3	8	10^{-1}	10^{-5}	[1,6]
3	8	10^{-1}	10^{-5}	[1,3]
3	8	10^{-1}	10^{-7}	$\varnothing$
3	8	10^{-1}	$1.1 \cdot 10^{-6}$	[4,6]
3	9	10^{-1}	10^{-6}	[1,6]
3	9	10^{-1}	10^{-6}	[1,4]
3	9	10^{-5}	10^{-6}	$\varnothing$
3	9	10^{-1}	$2.99 \cdot 10^{-7}$	[3,6]

Table 6.8 Results of problem (6.25) and (6.26) for the sample model with $N = 64$, $m = 5$

μ_n	μ_w	ε_n	ε_w	$[G_*, G^*]$
3	8	10^{-2}	10^{-4}	[1,2]
3	8	10^{-1}	10^{-4}	[1,4]
3	8	10^{-1}	$0.2 \cdot 10^{-4}$	[2,4]
3	14	10^{-2}	$7.45 \cdot 10^{-8}$	[2,4]
3	14	10^{-2}	10^{-9}	$\varnothing$
5	8	10^{-1}	10^{-5}	[1,4]
5	8	10^{-2}	10^{-6}	$\varnothing$
5	8	10^{-1}	$1.903 \cdot 10^{-5}$	[2,4]
5	14	10^{-1}	10^{-8}	$\varnothing$
9	14	10^{-2}	10^{-7}	[1,3]
9	14	10^{-1}	$6.643 \cdot 10^{-8}$	[2,5]

in them loading of w-calls significantly exceeds (by almost 10 times) loading of n-calls.

As the rate of change of function $PB_w(G)$ is low the results of the solution of the problem are very sensitive to possible changes in the upper limit (ε_w) of restriction (6.26). So, for example, at $N = 22$, $m = 2$, $\mu_n = 2$, $\mu_w = 9$, and $\varepsilon_n = 10^{-1}$ the problem has the optimal solution [1, 6] when $\varepsilon_w = 8.15 \cdot 10^{-7}$, and already with the same initial data the problem has no solution at $\varepsilon_w = 2.99 \cdot 10^{-7}$ (see Table 6.7). A similar situation occurs for the initial data indicated in Table 6.8. So, for example, for these initial data at $\mu_n = 3$, $\mu_w = 8$, $\varepsilon_n = 10^{-1}$ the problem has the optimal solution [1, 4] at $\varepsilon_w = 10^{-4}$ and already with the same initial data the problem has the optimal solution [2, 4] at $\varepsilon_w = 0.2 \cdot 10^{-4}$ (see Table 6.8).

The approach offered here allows the investigation of solutions of problem (6.25) and (6.26) at any admissible ratios of loadings of heterogeneous calls and at any possible values of structural parameters of the model. And each specific problem of choosing optimal values of threshold parameter G requires corresponding post-optimal analyses. At the same time, it is possible to draw the following general conclusions concerning properties of the solutions of problem (6.25) and (6.26):

- with increasing loading of w-calls the left end of an optimal interval (if it exists) moves to the right;
- with increasing loading of n-calls the right end of an optimal interval (if it exists) moves to the left;
- with decreasing ε_w the left end of an optimal interval (if it exists) moves to the right;
- with decreasing ε_n the right end of an optimal interval (if it exists) moves to the left;
- with simultaneously decreasing ε_n and ε_w the length of an optimal interval is reduced and below their certain values the problem has no solution;
- with increasing loadings of calls of any type the length of an optimal interval is reduced and above their certain critical values the problem has no solution.

We shall note that the problem of choice of an optimal value for the size of the buffer (R) can be similarly formulated and solved.

6.3 Problems for Mixed Models

In mixed models a very important QoS metric is the probability of degradation states. Therefore in these models, as a rule, there is a severe restriction on this parameter. On the other hand, channels are always in short-supply in terms of network resources. Differently, for the mixed models interest represents problems of optimization in which there are restrictions on these quantities.

In view of identifying a statement of problems and methods of their solution, only parametric optimization problems for models of mixed multi-rate systems with a discrete bandwidth for transfer of data (i.e. elastic) calls are considered here. To concretize a statement we shall consider the following problem.

Let restrictions on the upper limits of probabilities of degradation states of d-calls and blocking probabilities of heterogeneous calls be given (reminding ourselves, that for models with a discrete bandwidth of transfer the blocking probabilities of voice and data calls are equal to each other). Let's also remember that if the rate of servicing of a d-call is less than some threshold value δ, $\delta>0$, it indicates that the system is in a state of degradation (see Sect. 5.1).

So, the following restrictions are defined:

$$\mathrm{PB}(N) \le \varepsilon, \tag{6.27}$$

$$\mathrm{PDS}(N, \delta) \le \gamma, \tag{6.28}$$

where ε и γ are known upper limits for the indicated QoS metrics.

It is required to find such minimal number of channels of system (N^*) and the maximal value of threshold parameter (δ^*) for d-calls that the given restrictions on blocking probabilities of heterogeneous calls (6.27) and probabilities of degradation states (6.28) are satisfied.

Since the blocking probability does not depend on threshold parameter δ and both functions PB and PDS decrease monotonically with respect to argument N (for function *PDS* at the fixed value of threshold parameterδ), for the solution of problem (6.27) and (6.28) it is possible to offer the following algorithm.

First we need to find the minimal value N denoted through N_1, which satisfies restriction (6.27). To find N_1 the following scheme, in particular, can be applied. Limits of change of N_1 are defined proceeding from limits of change of function PB:

$$E_{\mathrm{B}}(\nu_v, N) \le \mathrm{PB}(N) \le E_{\mathrm{B}}(\nu_v + \nu_d, N). \tag{6.29}$$

The left limit in the inequality (6.29) directly follows from the formula for PB(N) (see step 2 of the algorithm in Sect. 5.1.2) and the right side of the limit is obvious.

After finding N_1 the following condition is checked: PDS(N_1, 1)$\le\gamma$. If this condition is satisfied, in an interval [1, N_1] find such maximal value δ_1 that condition PDS(N_1, δ_1)$\le\gamma$ is satisfied. If PDS(N_1, 1)$>\gamma$ then it is necessary to increase the number of channels until condition PDS(N_1, 1)$\le\gamma$ be satisfied and after that a corresponding value δ is determined.

Thus, summarizing it is possible to propose the following iterative algorithm of the solution of problem (6.27) and (6.28).

kth iteration.

Step 1. Find such minimal N_k that condition (6.27) was satisfied.
Step 2. If PDS(N_k,1)$\le\gamma$, then find such maximal value $\delta_k\in[1, N_k]$ that condition (6.28) was satisfied. Set $N^* := N_k$ and $\delta^* := \delta_k$.
Step 3. Set $N_{k+1} := N_k + 1$ and go to Step 1.

Table 6.9 Results of problem (6.27) and (6.28). $\nu_v = 10$, $\nu_d = 1$

ε	10^{-2}	10^{-2}	10^{-2}	10^{-2}	10^{-3}	10^{-3}	10^{-3}	10^{-3}	10^{-4}	10^{-4}	10^{-4}	10^{-4}	10^{-5}	10^{-6}	10^{-7}
γ	10^{-2}	10^{-3}	10^{-4}	10^{-5}	10^{-2}	10^{-3}	10^{-4}	10^{-5}	10^{-2}	10^{-3}	10^{-4}	10^{-5}	10^{-4}	10^{-4}	10^{-4}
N^*	18	19	22	25	21	21	22	25	24	24	24	25	27	29	31
δ^*	4	2	2	2	7	3	2	2	11	6	3	2	7	9	11

The algorithm is finite since function PDS (as has been noted above), is decreasing with respect to the number of channels at a fixed value of δ. Some results of the solution of this problem are shown in Table 6.9. The following conclusions are drawn from an analysis of the results of the problem:

- at fixed requirements to blocking probabilities (voice or data), stricter requirements for probability of degradation states leads to an increase of total number of necessary channels, but the corresponding maximal value of the threshold parameter for d-calls does not increase;
- at fixed requirements to probability of degradation states, stricter requirements for blocking probabilities of heterogeneous calls also leads to an increase in the total number of required channels, but this time the corresponding maximal value of the threshold parameter for d-calls also increases.

6.4 Conclusion

Problems related to improving MRQ characteristics by choice of appropriate parameters for the given access strategies are insufficiently investigated, although such problems have significant scientific and practical interest for the following reasons. First, application of the appropriate results allows sufficient improvement of the desired QoS metrics without particular effort. Second, it allows the study of all required QoS metrics of a system for the determined parameters of the access strategy and adapting them to the current traffic. Third, the determined access strategies are easily realized in real systems.

A few approaches towards parametric optimization of special models of MRQ are also known from the literature. For example, in [4] an algorithm for optimal partitioning of the total bandwidth among heterogeneous traffic through use of an integer programming technique is presented, where the objective of optimization is the minimization of the overall blocking probabilities. [1] considers the problem of finding an optimal CAC for MRQ with two types of calls, where the objective is the minimization of the weighted sum of blocking probabilities. Here the authors show that the optimal parametric strategy is described by two monotone switching curves, one for each call type. A similar model was considered in [3].

Parametric CAC based on the reservation mechanism is an effective way to improve the QoS metrics in multi-rate systems. A review of work on the asymptotic optimality of trunk reservation strategies can be found in [2]. One type of reservation strategy was considered in [5]. Therein the total bandwidth is divided into two parts: an active group that is available to all types of calls and a reserve group that is available in a controllable manner to call requests when at the moment of arrival the number of free channels in the active group is not sufficient to service this call.

Problems relating to improvement of MRQ characteristics by means of the parametric optimization problems examined in this chapter are investigated in [6–8].

References

1. Lambadaris I, Narayan P (1988) Optimal control of arrivals at a blocking node. In: Proceedings of computer networking symposium, Washington, DC, pp 209–213
2. Laws CN (1995) On trunk reservation in loss networks. In: Kelly FP, Williams KS (eds) Stochastic networks. Springer, New York, NY
3. Medhi D, Liefvoort A, Reece CS (1995) Performance analysis of a digital link with heteregeneous multislot traffic. IEEE Trans Commun 43:968–976
4. Meempat G, Sundareshan MK (1993) Optimal channel allocation policies for access control of circuit-switched traffic in ISDN environments. IEEE Trans Commun 41:338–350
5. Melikov AZ, Deniz DZ (1999) The optimal Markov strategy for access in ISDNs with reserves of channels. In: Proceedings of 7th IEEE Mediterranean conference on ion control and automation, Haifa, pp 2430–2437
6. Melikov AZ, Deniz DZ (2000) Non-exhaustive channel access strategy in multi-resource communication systems with non-homogeneous traffic. In: Proceedings of 5th IEEE symposium on computers and communications, France, pp 432–437
7. Melikov AZ, Fattakhova MI, Kaziyev TS (2006) Multiple-speed system with specialized channels for servicing broadband customers. Autom Control Comput Sci 40(2):11–19
8. Melikov AZ, Kim CS, Ponomarenko LA (2009) Algorithmic approach to studying models of multi-rate systems with queues. Cybern Syst Anal 45(1):76–83

Chapter 7
Markov Decision Processes (MDP) Approach to Optimization Problems for Multi-Rate Systems

In the previous chapter of the book various problems of improvement of characteristics of traditional and multimedia wireless communication networks were considered. At the same time improvement of desirable characteristics was reached, basically, due to a choice of corresponding values of parameters of the used access strategy, i.e. parametric optimization problems were examined. In other words, the admissible access strategy class is defined in advance, and the problem consists in definition of optimal (in a given sense) values of access strategy parameters.

Such an approach to the solving of the specified problems, naturally, restricts the set of strategies for which an optimum strategy of access can be determined. To analyze such problems over a wide class of strategies it is expedient to use the Markov Decision Processes (MDP) approach.

In this chapter both classical and new methods of MDP which allow the definition of optimal strategies of access in teletraffic systems are considered.

7.1 Hierarchical Phase-Merging Algorithm for MDP Problems

In multi-service teletraffic systems the calls differ from each other in relation to various parameters, for example, loading characteristics, importance, urgency, etc. This has led to the present activity in terms of research of models with various types of priorities.

Any priority discipline in teletraffic systems should define rules for (A) reception of a call in a system, (B) a choice of which following call is to be serviced after release of a channel, and (C) assignment of the channel. The need for definition of rule (C) arises in systems where channels are not identical (for example, differ from each other in terms of speed of service and/or in the cost of switching and working costs in a unit of time, etc.).

The above-mentioned rules A–C are realized by means of priorities of two types: exogenous and endogenous. Exogenous priorities for rules A–C are defined on the basis of rules developed in advance and do not take into account the current state of the system, and the state of the system can be determined in various ways. From the point of view of convenience the use of this type of priorities is preferable

L. Ponomarenko et al., *Performance Analysis and Optimization of Multi-Traffic on Communication Networks*, DOI 10.1007/978-3-642-15458-4_7,

since their realization does not demand any special expense in relation to software. However, real teletraffic systems operate under conditions of uncertainty regarding traffic parameters which makes it impossible in many cases for a fixed preliminary assignment of priorities. These circumstances have led to the present activity in terms of the study of problems of teletraffic systems with endogenous priorities.

Endogenous priorities in teletraffic systems can be determined by the following scheme. For each state the finite set of admissible decisions is defined, and for each decision some parameter estimating probabilities of acceptance of the given decision in a current state is introduced. Criteria of quality in concrete teletraffic systems are defined differently, proceeding from the purpose of the system. At the same time, researches of models with criterion evaluating total costs (economic, technical, etc.) resulting from the system sojourn in various states are very promising. In such cases Markov models represent a universal method for their calculation via its expression through stationary probabilities of states. This becomes possible due to the unique property of Markov systems in which the stationary probability of a state represents a part of the sojourn time of the system in a corresponding state for a large time interval.

Then the problem of definition of optimal priorities consists in a choice of corresponding decisions in conflict states, i.e. in states in which there is a necessity to choose a decision from some finite or infinite set of decisions. Thus, research of Markov models of teletraffic systems with endogenous priorities is equivalent to research of some MDP problems.

Let's formulate one of the possible ways to define the MDP problem. An undiscounted MDP problem with an infinite planning horizon is defined by the following objects.

1. The Markov chain with finite state space $X:=\{x_1, x_2, \ldots, x_N\}$ is specified (we shall use in the text following the symbol x_k for the notation of states, and in formulas for the notation of a state we will use the symbol k, $k=1,\ldots,N$).
2. The finite set of decisions D, $D:=\bigcup\limits_{k\in X} D_k$ is specified where D_k is the set of admissible decisions in a state $x_k\in X$.
3. A probability rule of decisions $\alpha_k{}^d:=\Pr\{\text{decision}=d \mid \text{state}=x_k\}$ and the corresponding transition matrix $P=\|\, p_{\mathrm{d}}\,(k, k')\,\|$, where $p_{\mathrm{d}}(k, k')$ is denoted the probability of transition from state x_k into state $x_k{}'$ at a choice of the state x_k decision (control) $\mathrm{d}\in D_k$, are defined. This leads to the following conditions being fulfilled:

$$\alpha_k^d \ge 0,\ \sum_{d\in D_k} \alpha_k{}^d = 1,\ \forall x_k \in X;$$

$$p_{\mathrm{d}}\left(k,k'\right) \ge 0,\ \sum_{k'\in X} p_{\mathrm{d}}\left(k,k'\right) = 1,\ \forall x_k \in X,\ \forall d \in D_k\,.$$

4. The vector of average costs for one step $\boldsymbol{C}=(C_1, C_2,\ldots,C_N)$ is defined as follows

$$C_k := \sum_{d\in D_k} c_k^d \alpha_k^d\,,\ x_k \in X.$$

where ${c_k}^d$ are costs for one step if in a state x_k decision $d \in D_k$ is accepted thus ${c_k}^d$ are uniformly restricted quantities. In other words, C_k are expected average costs connected with an output from a state x_k for one step, $k = 1, \ldots, N$.

The problem of optimization of the Markov Chain (MC) is formulated as follows: it is necessary to find such strategy of control to minimize average costs for one step:

$$W := \sum_{k \in X} C_k \, \overrightarrow{d \in D} \min . \qquad (7.1)$$

Problem (7.1) is solved, as a rule, subject to a system of balance equations for stationary probabilities $\rho(k)$, $x_k \in X$, by taking into account decisions $d \in D$. The corresponding problem of linear programming (LP) has the following form

$$W := \sum_{k,d} c_k^d x_k^d \to \min , \qquad (7.2)$$

subject to

$$\sum_{d} x_{k'}^d - \sum_{k,d} p_d\left(k, k'\right) x_k^d = 0, \; k' = 1, 2, \ldots, N , \qquad (7.3)$$

$$\sum_{k,d} x_k^d = 1, \; x_k^d \geq 0, \; k = 1, 2, \ldots, N; \; d = 1, 2, \ldots, M, \qquad (7.4)$$

where M is the dimension of set D.

It is well known that any optimal solution of problem (7.2), (7.3), and (7.4) has the following property: for each state x_k there is only $d = d(k)$, for which ${x_k}^{d(k)} > 0$ and all ${x_k}^d = 0$ at $d \neq d(k)$. This means that the optimal strategy is nonrandomized and does not depend on the initial distribution of a chain.

Problems of MDP are solved usually either by methods of dynamic (Dynamic Programming, DP) or linear programming (LP), thus in the sense of computational complexity both approaches are almost equivalent and lead to the same results. At the same time, for the optimization of real teletraffic systems it is suitable to use LP methods for two reasons. First, loadings of teletraffic systems are determined in practice with some error and consequently for designers of these systems the vital interest represents a question on in what ranges of change of loading parameters optimal strategies saves their kind. As is known, one can carry out extensive post-optimization analysis with modern LP software packages, in particular, to respond to the question specified above. Secondly, when using LP there is an opportunity to consider some additional non-linear constraints (though it is then not guaranteed to be a non-randomized property of the determined strategy). Nevertheless, the application of DP does not allow one to resolve these problems.

We shall name the approach described above to research MDP problems exact. This approach appears to be effective in the research of MDP problems for which the state space of the initial controlled MC contains a small number of microstates.

It is obvious that the state space of models of teletraffic systems with significant traffic and large values for system structural parameters (i.e. number of channels and buffer dimension) contain a huge quantity of microstates. Therefore, using the exact approach (EA) for optimization of large teletraffic systems encounters significant computing difficulties which are not easily solved even with modern computers. Below we present a problem for simplification of the description of such systems.

Here a new hierarchical algorithm (approximate) of the space-merging type which simultaneously uses a principle of decomposition is proposed and does not impose any restrictions on the structure of the generating matrix (GM) of the corresponding controlled MC.

The hierarchical algorithm developed here allows the solving of problems of optimization of controlled MC in practically any dimension. Each step of the hierarchy consists of identical steps therefore for reasons of simplification of notation we shall describe the work of the algorithm only at the first step.

Let a finite undiscounted MDP with an infinite planning horizon be defined by means of objects 1–4 as specified above.

Step 1. Some splitting state space X (see Fig. 7.1a) is considered:

$$X = \bigcup_{v=1}^{V} X_v \, , \, X_v \cap X_{v'} = \varnothing \, , \, v \neq v'. \tag{7.5}$$

Step 2. All the microstates in subset X_v are combined into one merged state denoted ω_v, $v = 1,2,\ldots,V$. All states obtained in this way form a space $\Omega := \{ \omega_1, \omega_2, \ldots, \omega_V \}$.

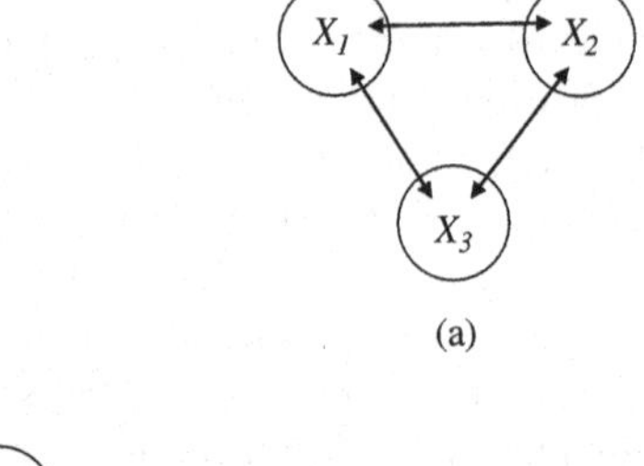

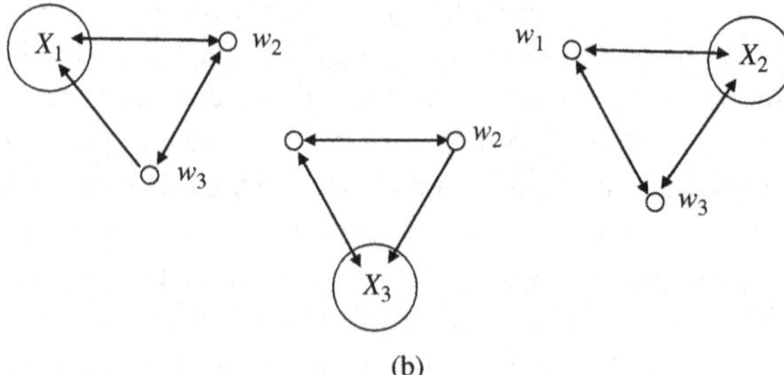

Fig. 7.1 Splitting of state space X of initial model (**a**) and construction of state space of merged models (**b**), V=3

Step 3. On the basis of splitting (7.5) merge functions $U_v : X \to \hat{X}_v$are constructed where$\hat{X}_v = X \bigcup (\Omega - \{\omega_v\})$, $v = 1, 2, \ldots, V$. These functions are defined as follows:

$$U_v(x) = \begin{cases} x, & \text{if} \, x \in X_v \\ \omega_{v'}, & \text{if} \, x \in X_{v'}, \; v' \neq v \end{cases}. \tag{7.6}$$

Merge functions (7.6) define V merged (with respect to the initial model) models, thus the vth merged model has state space $\hat{X}_v$, v=1,2,...,V. Various variants of construction of the merged models are represented in Fig. 7.1b.

Step 4. The set of admissible decisions D^v for the vth merged model is determined. It is determined as a projection of set D to set X_v, i.e. $D^v := \bigcup\limits_{x \in X_v} D_x$.

Step 5. Elements of a transition matrix of the vth merged model $P_v = ||P_v{}^d(x,x')||$, x, $x' \in \hat{X}_v$ are defined:

$$P_v^d(x,x') = \begin{cases} p^d(k,k')\rho(k), & \text{if } x = x_k, x' = x_k{}' \\ \sum\limits_{k' \in X_v{}'} p^d(k,k')\rho(k), & \text{if } x = x_k, x' = \omega_k{}' \\ \sum\limits_{k' \in X_v{}'} \sum\limits_{d \in D_k} p^d(k,k')\rho(k), & \text{if } x = \omega_v{}', x' = x_k{}' \\ \sum\limits_{\substack{k \in X_v{}' \\ k' \in X_v{}''}} \sum\limits_{d \in D_k} p^d(k,k')\rho(k), & \text{if } x = \omega_v{}', x' = \omega_v{}'' \end{cases} \tag{7.7}$$

As the stationary distribution of the initial model is not known, on researching the vth merged model it is impossible to use values $P_v{}^d\left(x, x'\right)$, $d \in D^v$ determined by means of formula (7.7). Hence, it is necessary, using various schemes of approximation to estimate values of unknown elements of a transition matrix from above or from below.

Note 7.1. From a practical point of view such estimation is expediently made from above as in this way the final results will be more reliable. In particular, for such estimation the following fact can be used:

$$\text{if } a_i \geq 0, \;\; i = 1,2, \ldots, K \text{ and } 0 < \sum_{i=1}^{K} b_i \leq 1 \text{ then } \sum_{i=1}^{K} a_i b_i \leq \max_i \{a_i\}.$$

Step 6. Elements of a vector of average costs are determined through the unknown stationary distribution of the initial model. Let for the vth merged model ${}^vC^d{}_x$ mean costs for one step if in a state $x \in \hat{X}_v$ decision $d \in D^v$ is accepted. Therefore, here too there is a necessity for approximation of values of these quantities and the estimations for cases from above and from below are defined by formulas (7.8) and (7.9), respectively.

$$
{}^{v}\overline{C}_{x}^{d} = \begin{cases} C_{k}^{d}, & \text{if } x = x_k \in \hat{X}_v, \\ \max\limits_{d \in D_k} \max\limits_{k \in X'_v}(C_{k}^{d}), & \text{if } x = \omega'_v \in \hat{X}_v, v' \neq v; \end{cases} \quad (7.8)
$$

$$
{}^{v}\underline{C}_{x}^{d} = \begin{cases} C_{k}^{d}, & \text{if } x = x_k \in \hat{X}_v, \\ \min\limits_{d \in D_k} \min\limits_{k \in X'_v}(C_{k}^{d}), & \text{if } x = \omega'_v \in \hat{X}_v, v' \neq v. \end{cases} \quad (7.9)
$$

Step 7. For approximation of criterion (7.1) estimations from above and from below are used, determined by formulas (7.10) and (7.11.), respectively.

$$
{}^{v}\overline{W} := \sum_{x \in \hat{X}_v} {}^{v}\overline{C}(x), \quad (7.10)
$$

$$
{}^{v}\underline{W} := \sum_{x \in \hat{X}_v} {}^{v}\underline{C}(x), \quad (7.11)
$$

where

$$
{}^{v}\overline{C}(x) := \sum_{x \in D^v} {}^{v}\overline{C}_{x}^{d} \alpha_{x}^{d}; \; {}^{v}\underline{C}(x) := \sum_{x \in D^v} {}^{v}\underline{C}_{x}^{d} \alpha_{x}^{d}.
$$

Step 8. If in all merged models all parameters (elements of a transition matrix and criteria) are estimated from above we shall receive V majority problems with respect to the initial optimization problem. And if in all merged models the mentioned parameters are estimated from below we shall receive V minority problems with respect to the initial optimization problem. As a result of the solution of the problem of optimization of the vth merged model (majority or minority) there are approximate optimal values α_k^d, $x_k \in X_v$, and, thus, after the parallel solution of all similar V problems there are approximately optimal values of all α_k^d where $x_k \in X$. The accuracy of the proposed method is estimated in the following way:

$$
\max_v {}^{v}\underline{W^*} \leq W^* \leq \min_v {}^{v}\overline{W^*} \quad (7.12)
$$

where W^*, $\underline{W^*}$, $\overline{W}^*$ are optimal values of the criterion in the initial, vth majority, and vth minority problem, respectively, $v = 1, 2, \ldots, V$.

The important advantage of the developed approach is that unlike known approaches it does not impose any restrictions on the structure of the transition matrix of the initial model. Moreover, it provides an opportunity for repeated application for construction of merged models, and, thus, enables hierarchies to be built up in problems of optimization of controlled MC. The latter means, that if the dimension of the optimization problem of the initial model of a large-scale MC is not reduced as a result of a single application of the developed algorithm it is necessary to repeatedly apply the given algorithm to the merged models and then the hierarchy of the merged models becomes apparent.

The accuracy of the proposed approach is influenced by two factors: the scheme of splitting of state space of the initial model and the scheme of approximation of unknown parameters in the merged models.

The merged models are defined by the choice of a concrete scheme of splitting of state space of the initial model, and, hence, it is possible to use various schemes of splitting. Concerning estimation of unknown parameters in the merged models it is necessary to note that if more careful schemes of approximation are used the results are closer to the exact solution. Thus, estimation (7.12) remains correct in all cases. These facts confirm the results of the numerical experiments in relation to the application of a non-hierarchical variant of the given algorithm for optimization of queuing systems with priorities.

It is necessary to note, that the question of whether the optimal values of controls will be identical for the exact and approximate approaches remains open.

7.2 Finding the Optimal Access Strategy

In the previous chapters models of multi-rate queues were investigated for various access strategies. Therefore, here we shall not dwell on detailed descriptions of their operation. We shall note that all notations of the previous chapters related to such models are used without change.

As was mentioned in Sect. 7.1, the choice of a criterion of the system depends on its concrete goal. To make a definite statement here it is supposed that QoS of the general model of un-buffered MRQ is estimated by the average number of busy channels. The problem consists in the following: it is required to find such strategy of access at which it would be possible to maximize the loading of channels of the mentioned system. In other words, the purpose is the solution of the following problem:

$$\tilde{N} \to \max, \tag{7.13}$$

where $\tilde{N} := \sum_{n \in S} (\mathbf{n},\mathbf{b})\, p(\mathbf{n})$ is an average number of busy channels, and S denotes the state space of the model which was defined in Sect. 4.1.1.

The strategy of access, being the solution of problem (7.13), is called optimal. Here it is necessary to note that optimal call admission control (CAC) is looked for in a class of strategies which does not suppose interruptions of already begun service processes of calls of any type.

Should a call of any type arrive in the system at a moment when the necessary number of channels are not free then it is supposed that with probability 1 it is rejected. Hence optimal CAC should estimate the chance of arrived calls in those moments when there are a necessary number of free channels. In fact at these moments two alternative decisions are possible: (1) the newly arrived call is accepted for service; (2) it is rejected.

With the purpose of definition of controllable decisions we shall examine the moment of arrival of a call of ith type, i=1,2, ..., K where K indicates the total number of types of heterogeneous calls. Let at this moment the system be in a state $\mathbf{n}\in S$ in which $f(\mathbf{n})\geq b_i$ where $f(\mathbf{n}):=N-(\mathbf{n},\mathbf{b})$ means the number of free channels in a given state; otherwise (i.e. when $f(\mathbf{n})<b_i$) as was specified above, the call of ith type is rejected. Then acceptance of one of two decisions is possible: (a) the call of the ith type is accepted into service, or (b) it is rejected.

Probabilities of acceptance of decisions (a) and (b) are denoted by $\alpha_i^+(\mathbf{n})$ and $\alpha_i^-(\mathbf{n})$, respectively. These probabilities are called controllable decisions. They satisfy the following conditions:

$$0 \leq a_i^+(\mathbf{n}) \leq 1; \tag{7.14}$$

$$a_i^+(\mathbf{n}) + a_i^-(\mathbf{n}) = 1, \forall\, i \in F(\mathbf{n}), \tag{7.15}$$

where

$$F(\mathbf{n}) = \left\{i \in Z_k{}^+ : f(\mathbf{n}) \geq b_i\right\}, Z_K{}^+ := \{1, \ldots, K\}.$$

Note 7.2. At specific values of the introduced controllable decisions known strategies are obtained. So, if $\alpha_i^+(\mathbf{n}) = 1$ for any $i\in F(\mathbf{n})$ the CS-strategy of access is obtained; if $\alpha_i^+(\mathbf{n}) = 1$ for any $i\in Z_K^+$ under condition $f(\mathbf{n})\geq b$ where $b := \max\{b_i\}$ the CSE-strategy of access is obtained; if $\alpha_i^+(\mathbf{n}) = 1$ for any $i\in Z_K^+$ under condition $f(\mathbf{n}) \geq b_i+r_i$, $0 \leq r_i \leq N-b_i$ the TR-strategy of access is obtained. Other CAC are also possible at certain values of controllable decisions in the above-described scheme. Hence the described scheme defines a wide class of access strategies in which the push-out of calls from the channel is not allowed.

On using the introduced controllable decisions the elements of GM $q(\mathbf{n},\mathbf{n}')$, $\mathbf{n}$, $\mathbf{n}'\in S$, are determined thus:

$$q(\mathbf{n}, \mathbf{n}') = \begin{cases} \lambda_i\alpha_i^+(\mathbf{n}), & \text{if} \quad \mathbf{n}' = \mathbf{n} + \mathbf{e_i}, \\ n_i\mu_i, & \text{if} \quad \mathbf{n}' = \mathbf{n} - \mathbf{e_i}, \\ 0 \ \text{in other cases}. \end{cases} \tag{7.16}$$

Hence, in an explicit form problem (7.13) can be written as follows:

$$\sum_{n\in S} (\mathbf{n}, \mathbf{b})\, p(\mathbf{n}) \to \max \tag{7.17}$$

subject to

$$\begin{aligned} &\sum_{i=1}^{K} \left(\lambda_i\alpha_i^+(\mathbf{n}) I\left(f(\mathbf{n}) \geq b_i\right) + n_i\mu_i\right) p(\mathbf{n}) = \sum_{i=1}^{K} \lambda_i p(\mathbf{n} - \mathbf{e_i})\alpha_i^+(\mathbf{n} - \mathbf{e_i})\chi(n_i) \\ &\quad + \sum_{i=1}^{K} (n_i + 1)\,\mu_i p\,(\mathbf{n} + \mathbf{e_i})\, I(f(\mathbf{n}) \geq b_i), \ \ \mathbf{n} \in S; \end{aligned} \tag{7.18}$$

$$\sum_{n \in S} p(\mathbf{n}) = 1; \tag{7.19}$$

$$\alpha_i^+(\mathbf{n}) + \alpha_i^-(\mathbf{n}) = 1, \ \forall i \in F(\mathbf{n}), \mathbf{n} \in S. \tag{7.20}$$

In problem (7.17), (7.18), (7.19), and (7.20) constraints (7.18) and (7.19) set balance equations for the given model on use of controllable decisions and they are composed in view of (7.16). This is a MDP problem and as noted in Sect. 7.1, has an optimal solution according to which the values $\alpha^{\pm}{}_i(\mathbf{n})$ are equal either to 0, or 1 for each $i \in F(\mathbf{n})$. The latter circumstance allows construction of a simple algorithm for realizing the determined optimal non-randomized CAC in the investigated system.

Note 7.3. As a result of the problem solution, along with optimal values of CSP $\alpha_i{}^{\pm}(\mathbf{n}), \mathbf{n} \in S, i \in F(\mathbf{n})$, also found is the stationary distribution of the system $p(\mathbf{n})$, $\mathbf{n} \in S$. The latter permits us also to calculate stationary blocking probabilities of heterogeneous calls. Thus, blocking probability of i-calls (PB_i) is calculated in the following way:

$$\mathrm{PB}_i = \sum_{f(n)=0}^{b_i-1} p(\mathbf{n}) + \sum_{f(n)=b_i}^{N} p(\mathbf{n})\alpha_i{}^-(\mathbf{n}), \ i = 1, 2, \ldots, K. \tag{7.21}$$

The first sum in the formula (7.21) calculates losses related to blocking of i-calls, those that appear with probability of 1 when inequality $b_i > f(\mathbf{n})$ holds true; and the second sum calculates losses caused by blocking of i-calls as a result of rejecting them in handling that occurs with the probability of $\alpha_i{}^-(\mathbf{n})$ when inequality $b_i \leq f(\mathbf{n})$ holds true.

The EA described above to the solution of the problem of finding the optimal CAC is effective at small values of N and K, but with their growth dimension S grows exponentially (see Sect. 4.1). In this connection use of the EA becomes practically impossible. For the solution of this problem for large-scale MRQ it is proposed below that we use the approximate approach developed in Sect. 7.1.

Let's consider the following splitting S:

$$S = \bigcup_{r=0}^{N} S_r, \ S_r \bigcap S_{r'} = \varnothing, \ r \neq r', \tag{7.22}$$

where $S_r := \{\mathbf{n} \in S\colon (\mathbf{n},\mathbf{b}) = r\}$, i.e. class S_r includes those states $\mathbf{n} \in S$ in which the total number of busy channels equals r. Furthermore, each class S_r is described by one merged state denoted through $<r>$, $r = 0, \ldots, N$. On the basis of splitting (7.22) the following merge functions are constructed:

$$U_r : S \to \hat{S}_r, \ \hat{S}_r := S_r \bigcup \left(Z_N^0 - \{<r>\}\right), \ Z_N^0 = \{0, 1, \ldots, N\}. \tag{7.23}$$

Merge functions (7.23) are defined thus:

$$U_r(\mathbf{n}) = \begin{cases} \mathbf{n}, & \text{если} \quad \mathbf{n} \in S_r, \\ < r' >, & \text{если} \quad \mathbf{n} \in S_{r'},\ r \neq r'. \end{cases} \tag{7.24}$$

Hence, merge functions (7.24) define the N+1 merged model, thus the rth merged model has state space $\hat{S}_r$. As all the integrated models have a similar structure below we shall fix value $r \in Z_N{}^0$ and we shall consider the rth model.

Let's pass to definition of the elements the GM of the rth merged model denoted $q_r(x,y)$, x, $y \in \hat{S}_r$. According to (7.16) we have:

$$q_r\left(\mathbf{n}, \mathbf{n}'\right) = 0,\ \forall \mathbf{n}, \mathbf{n}' \in \hat{S}_r,\ \mathbf{n} \neq \mathbf{n}'; \tag{7.25}$$

$$q_r(\mathbf{n}, < r' >) = \begin{cases} \sum\limits_{j=1}^{K} \lambda_j \alpha_j^{+}(\mathbf{n}) p(\mathbf{n}) \delta(b_j, b_i), & \text{if } r' = r + b_i, i = \overline{1, K}, \\ \sum\limits_{j=1}^{K} n_j \mu_j p(\mathbf{n}) \delta(b_j, b_i), & \text{if } r' = r - b_i,\ b_i \leq r, i = \overline{1, K}, \\ 0, & \text{in other cases}; \end{cases} \tag{7.26}$$

$$q_r(< r' >, \mathbf{n}) =$$
$$\begin{cases} \sum\limits_{j=1}^{K} \sum\limits_{\mathbf{n}-\mathbf{e}_j \in S_{r'}} \lambda_j \alpha_j^{+}(\mathbf{n} - \mathbf{e_j}) p(\mathbf{n} - \mathbf{e_j}) \chi(n_j) \delta(b_j, b_i), & \text{if } r' = r - b_i, i = \overline{1, K}, \\ \sum\limits_{j=1}^{K} \sum\limits_{\mathbf{n}+\mathbf{e_j} \in S_{r'}} (n_j + 1) \mu_j p\left(\mathbf{n} + \mathbf{e_j}\right) \delta\left(b_j, b_i\right), & \text{if } r' = r + b_i,\ i = \overline{1, K}, \\ 0, & \text{in other case}; \end{cases} \tag{7.27}$$

$$q_r(< r' >, < r'' >) =$$
$$\begin{cases} \sum\limits_{j=1}^{k} \sum\limits_{\mathbf{n}' \in S_{r'}} \lambda_j \alpha_j^{+}(\mathbf{n}') p(\mathbf{n}') \chi(\mathbf{n_j}) I(N - r' \geq b_j) \delta(b_j, b_i), & \text{if } r'' = r' + b_i, \\ \sum\limits_{j=1}^{K} \sum\limits_{\mathbf{n} \in S_{r'}} n_j' \mu_j p(\mathbf{n}') \delta(b_j, b_i), & \text{if } r'' = r' - b_i,\ b_i \leq r, i = \overline{1, K}, \\ 0, & \text{in other case}; \end{cases} \tag{7.28}$$

where $\mathbf{n}' = \left(n'_1, n'_2, \ldots, n'_K\right)$.

The validity of expressions (7.25), (7.26), (7.27), and (7.28) can be proved as follows. We shall choose some state $\mathbf{n} \in \hat{S}_r$. As transitions between states are possible only at the point calls enter and at the moment of termination of their service we shall consider these moments separately. After the termination of service of a call of the ith type in a state $\mathbf{n} \in \hat{S}_r$ there is a transition $\mathbf{n} \rightarrow \mathbf{n} - \mathbf{e}_i$ where $\mathbf{n} - \mathbf{e_i} \in S_{r-b_i}$(i.e. $\mathbf{n} - \mathbf{e_i} \notin S_r$).

If at the moment a call of the ith type enters a system the system is in a state $\mathbf{n} \in \hat{S}_r$ in which $f(\mathbf{n})<b_i$ [i.e. $i \notin F(\mathbf{n})$], the newly arrived call with probability 1 is lost; otherwise (i.e. when $f(\mathbf{n}) \geq b_i$) two decisions are possible: (1) if the call of ith type is accepted [with probability $\alpha_i^+(\mathbf{n})$] there is a transition $\mathbf{n} \rightarrow \mathbf{n}+\mathbf{e}_i$ where $\mathbf{n}+\mathbf{e_i} \in S_{r+b_i}$(i.e. $\mathbf{n}+\mathbf{e}_i \notin S_r$); (2) if a call of ith type is rejected [with probability $\alpha_i^-(\mathbf{n})$] there is a virtual transition $\mathbf{n} \rightarrow \mathbf{n}$.

Hence, in any state $\mathbf{n} \in \hat{S}_r$, $r = 0,1,\ldots, N$, after the termination of service of a call of ith type the system passes into state $\mathbf{n}-\mathbf{e_i} \in S_{r-b_i}$, and at the moment a call of ith type enters the system upon acceptance of various decisions there occurs in the system either a virtual transition (i.e. it remains in the same state) or the system passes into state $\mathbf{n}+\mathbf{e_i} \in S_{r+b_i}$, in other words, equality (7.25) is correct.

Let's determine $q_r\left(\mathbf{n}, <r'>\right)$, $\mathbf{n}, <r'> \in \hat{S}_r$. From a micro-state $\mathbf{n} \in \hat{S}_r$ a transition to merged state $<r'>$ is possible only when $r' = r+b_i$ if $r \leq N-b_i$ or $r' = r-b_i$ if $n_i > 0$. Transition $\mathbf{n} \rightarrow r+b_i$ can occur at the moments of entering into the system of a call of ith type in a state $\mathbf{n}$, where $f(\mathbf{n}) \geq b_i$ [i.e. $i \in F(\mathbf{n})$] since in this case the newly entered call with probability $\alpha_i^+(\mathbf{n})$ is accepted and, hence, there is a transition $\mathbf{n} \rightarrow \mathbf{n}+\mathbf{e}_i$ where $\mathbf{n}+\mathbf{e_i} \in S_{r+b_i}$ (i.e. there is a transition into the merged state $<r+b_i>$). After the termination of service of the call of ith type in a state $\mathbf{n} \in S_r$ there is a transition $\mathbf{n} \rightarrow \mathbf{n}-\mathbf{e}_i$ where $\mathbf{n}-\mathbf{e_i} \in S_{r-b_i}$ (i.e. there is a transition into the merged state $<r-b_i>$). Thus, for calculation $q_r\left(\mathbf{n}, <r'>\right)$, $\mathbf{n}, <r'> \in \hat{S}_r$ in view of the theorem of total probability, we shall receive equality (7.26).

By similar argument, we find that transition $<r'> \rightarrow \mathbf{n}, <r'>, \mathbf{n} \in \hat{S}_r$ is possible only in cases $r' = r-b_i$ or $r' = r+b_i$ and transition $<r'> \rightarrow <r''>$, $<r'>, <r''> \in \hat{S}_r$ is possible only in cases $r''' = r'+b_i$ or $r''' = r'-b_i$. The intensity of transitions between these states is defined from equations (7.27) and (7.28), respectively. Thus, the validity of expressions (7.25), (7.26), (7.27), and (7.28) is proved.

For construction of balance equations the given merged model should approximate exact values $q_r(x,y), x, y \in \hat{S}_r$ in expressions (7.26), (7.27), and (7.28). This is evident from the fact that these expressions include the stationary distribution of the initial model, and also controllable decisions which are not defined for the given merged model.

With this purpose in mind we shall use *Note* 7.1 of the previous section. Then quantities $q_r(x,y), x, y \in \hat{S}_r$ are estimated from above as follows:

$$\sum_{i=1}^{K} \lambda_i \alpha_i^+(\mathbf{n}) p(\mathbf{n}) \delta\left(r+b_i, r'\right) \leq \sum_{i=1}^{K} \lambda_i \alpha_i^+(\mathbf{n}) \delta\left(r+b_i, r'\right), \tag{7.29}$$

$$\sum_{i=1}^{K} n_i \mu_i p(\mathbf{n}) \delta\left(r-b_i, r'\right) \leq \max_i \left\{n_i \mu_i : r-b_i = r'\right\}, \tag{7.30}$$

$$\sum_{i=1}^{K} \sum_{n-e_i \in S_{r'}} \lambda_i \alpha_i^+(\mathbf{n}-\mathbf{e_i}) p(\mathbf{n}-\mathbf{e_i}) \chi(n_i) \leq \max_i \{\lambda_i\}, \tag{7.31}$$

$$\sum_{i=1}^{K}\sum_{n+e_i\in S_{r'}}(n_i+1)\mu_i p\left(\mathbf{n}+\mathbf{e_i}\right)\delta\left(r+b_i,r'\right)\leq \max_i\left\{(n_i+1)\,\mu_i : r+b_i=r'\right\}, \tag{7.32}$$

$$\sum_{i=1}^{K}\sum_{n'\in S_{r'}}\lambda_i\alpha_i^{+}\left(\mathbf{n}'\right)p\left(\mathbf{n}'\right)I\left(N-r'\geq b_i\right)\leq \max_i\left\{\lambda_i : b_i\leq N-r'\right\}, \tag{7.33}$$

$$\sum_{i=1}^{K}\sum_{n'\in S_{r'}}n_i'\mu_i p\left(\mathbf{n}'\right)\leq \max_i\left\{n_i'\mu_i : r''=r'-b_i\right\}. \tag{7.34}$$

Use of inequality (7.29) for an estimation $q_r(\mathbf{n},r')$ in the case of $r'=r+b_i$ in such form is explained by the fact that unknown controllable decisions $\alpha^{\pm}{}_i(\mathbf{n})$ participate on the left side of this inequality which should be determined as a result of the solution of a problem of optimization of the rth merged model. It is necessary to note also, that on the right side of inequalities (7.29), (7.30), (7.31), (7.32), (7.33), and (7.34) more rough estimations (and at the same time simpler) can be used.

Thus, as the approximate values of $q_r\left(x,y\right), x,y\in\hat{S}_r$ in (2.15), (2.16), (2.17), (2.18), (2.19), and (2.20) we use the following expressions:

$$q_r\left(\mathbf{n},r'\right)\begin{cases}\approx \sum_{i=1}^{K}\lambda_i\alpha_i^{+}(\mathbf{n})\delta(r+b_i,r') & \text{if } r'=r+b_i,\\ \approx \max_i\left\{n_i\mu_i : r-b_i=r'\right\} & \text{if } r'=r-b_i,\\ =0, \text{ in other cases;}\end{cases} \tag{7.35}$$

$$q_r(r',\mathbf{n})\begin{cases}\approx \max_{\mathrm{i}}\{\lambda_i : n_i>0\} & \text{if } r'=r-b_i,\\ \approx \max_i\left\{(n_i+1)\mu_i : r+b_i=r'\right\} & \text{if } r'=r+b_i,\\ =0, \text{ in other cases;}\end{cases} \tag{7.36}$$

$$q_r(r',r''')\begin{cases}\approx \max_{\mathrm{i}}\left\{\lambda_i : b_i\leq N-r'\right\} & \text{if } r''=r'+b_i,\\ \approx \max_i\left\{n_i\mu_i : (\mathbf{n}',\mathbf{b})=r', r''=r'-b_i\right\} & \text{if } r''=r'-b_i,\\ =0 \text{ in other cases.}\end{cases} \tag{7.37}$$

Balance equations for stationary probabilities of states $\pi\left(x\right), x\in\hat{S}_r$ of the rth merged model are composed on the basis of expressions (7.25), (7.35), (7.36), and

(7.37). In view of the evidence of its drawing up the explicit form of this balance equation is not shown here.

The value of the functional in the rth merged model is approximated from above as follows:

$$\tilde{N}\left(\hat{S}_r\right) := \sum_{n \in \hat{S}_r} (\mathbf{n}, \mathbf{b})\, \pi(\mathbf{n}) + \sum_{r' \in \hat{S}_r} r' \pi\left(<r'>\right). \tag{7.38}$$

Hence the problem of optimization of the rth merged model consists in maximization of the functional (7.38). Therefore, constraints of the given problem are balance equations for $\pi\ (x), x \in \hat{S}_r$ and constraints (7.20) which have been written down only for micro-states $\mathbf{n} \in S_r$.

The given problem as well as the problem of optimization of the initial model concerns a class of MDP problem. Hence, as a result of its solution there are approximately optimal values of controllable decisions $\alpha^{\pm}(\mathbf{n})$ where $\mathbf{n} \in S$, $i \in F(\mathbf{n})$. Thus, as has been noted in the previous section [see (7.12)]

$$\tilde{N}^*\ (S) \leq \min_{r \in Z_N^0} \tilde{N}^*\left(\hat{S}_r\right),$$

where $\tilde{N}^*(S)$ is the optimal (maximal) value of the criterion in the initial problem (2.5), (2.6), (2.7), and (2.8); $\tilde{N}^*\left(\hat{S}_r\right)$ is the optimal value of criterion (7.38) in the rth problem, $r \in Z_N^0$.

Let's consider some remarks concerning dimensions of problems of optimization of initial and merged models. It is known that the dimension of optimization problems corresponds to the dimension of state space of the investigated model. On use of splitting scheme (3.22) the dimension of state space of the rth merged model is equal to $\left|\hat{S}_r\right| = |S_r| + N$ where $|X\,|$ means the dimension of set X. From here it is clear that the dimension of state space $\hat{S}_r$will be much less, than the dimension of state space of the initial model S. So, for example, for model MRQ with parameters N=10, K =10, $\mathbf{b} = (1,2,\ldots,10)$ the dimension of state space of S is equal to 139, and state space $\hat{S}_9$ having the maximal dimension among all merged models, contains 35 elements, i.e. application of the approximate approach allows one to reduce the dimension of an initial problem almost four times.

Here we shall note that if for super-large MRQ as a result of a single application of a procedure of merging it does not sufficiently reduce the dimension of a problem of optimization of the initial model it is necessary to repeatedly apply the already applied procedure of merging and then the hierarchy of the merged models (see Sect. 7.1) is formed.

It is important to note also, that on construction of the merged models the scheme of splitting the state space of the initial model (2.9) is not unique. So, for example, it is possible to also use the following splitting of the mentioned state space:

$$S = \bigcup_{v=0}^{V} S^{v}, \; S^{v} \bigcap S^{v'} = \varnothing, \; v \neq v', \tag{7.39}$$

where $S^{v} := \left\{ \mathbf{n} \in S : \sum_{i=1}^{K} n_i = v \right\}$, $V := [N/b_i]$. In other words, class S^v includes those states $\mathbf{n} \in S$ in which the total number of heterogeneous calls equals v. The problem of construction of the merged models on use of splitting (7.39) is given to the reader.

Let's consider results of numerical experiments for model MRQ with parameters $N = 10$, $K = 10$, $\mathbf{b} = (1,2,\ldots, 10)$. As has been specified above, the state space of the given model for the exact approach contains 139 microstates. The problem of maximization of loading of channels in the MRQ was solved with the following values of loading parameters: (1) $(\lambda_1, \ldots, \lambda_{10}) = (0.25, \ldots, 0.25)$; (2) $(\lambda_1, \ldots, \lambda_{10}) = (\underbrace{0.25, \ldots, 0.25}_{5}, 0.5, \ldots, 0.5)$; (3) $(\lambda_1, \ldots, \lambda_{10}) = (\underbrace{0.25, \ldots, 0.25}_{5}, 1, \ldots, 1)$.

In all variants it is accepted that $(\mu_1, \ldots, \mu_{10}) = (1, \ldots, 1)$. The results of the numerical experiments are shown in Table 7.1. In column CS the values of loading of channels are specified under the CS-strategy of access (i.e. in the case of $\alpha^{+}(\mathbf{n}) = 1, \forall\, i \in F(\mathbf{n}), \mathbf{n} \in S$); in column EA the optimal (maximal) values of loading of the channels from the exact approach are specified; in column AA (Approximate Approach) the top lines (the bottom lines) correspond to optimal values of loading of the channels from the approximated approach using splitting (7.22) and (7.39).

The main goal of the solution of a CAC optimization problem consists in finding of optimal values of controllable decisions $\alpha_i^{\pm}(\mathbf{n})$. In this connection it is necessary to note, that for the investigated model of MRQ optimal values of these parameters are identical for both the exact and approximated approaches. At the same time, as noted in Sect. 1.2, generally it is impossible to prove, that an optimal CAC, determined by means of various approaches, will coincide.

Numerous experiments have shown, that introduction of an optimal CAC allows utilization of channels to be scientifically improved, thus the effect of improvement depends on loading parameters of heterogeneous traffic. Numerical experiments also have shown that generally the optimal CAC does not belong to a

Table 7.1 Results for optimization of MRQ using various approaches

	Channels utilization		
Number of variant	CS	EA	AA
1	1.176	3.214	3.856 3.452
2	3.029	5.738	6.043 5.988
3	7.829	8.945	9.056 9.112

class of strategy of the threshold type according to which narrow-band calls are accepted only when the number of free channels is less than the certain threshold value.

7.3 Finding the Sub-Optimal Access Strategy

In practice, especially while researching general MRQ models with a relatively large number of types of calls, the system state isn't observed completely. Only partial information about the system state is available, namely, only the general number of busy (free) channels is observed. Therefore, the desired CAC has to make a decision based on limited information. Let us name this optimal (in the definite meaning) CAC based only on information about the number of busy (free) channels *suboptimal*. It should be noted that in most cases, the main interest is not in finding the optimal, but the *sub-optimal access strategy*. This is explained by the fact, that in practice, traffic parameters are usually calculated with some inaccuracy, therefore the effect from use of optimal CAC usually is not very significant.

In this section a method is proposed for calculation of the sub-optimal CAC in a general MRQ model with pure losses. It is based on the hierarchical phase-merging algorithm for MDP problems (see Sect. 7.1).

It should be noted, based on the latter statement that the problem of calculating the sub-optimal CAC does not differ from that of calculating the optimal one. At the same time, as is shown below, controlling parameters, which evaluate the probability of making an alternative decision in described problems, differ from one another.

The exact approach to solution of the problem of calculation of optimal CAC is effective with little values of N and K, but should they increase, the dimension of S increases exponentially. Thus, for example, under the condition $N=10$, $\mathbf{b}=(1,2,3,\ldots,10)$ the dimension of S is 139, but when $N=30$, $\mathbf{b}=(1,2,3,\ldots,30)$ this value is 28628. In connection with this, the use of the exact approach becomes practically impossible. Therefore, the problem of calculation of a sub-optimal CAC for MRQ that is practically relevant for a system of large dimension is described below.

Since, according to the assumption, only the number of free (busy) channels are observed, let us examine the splitting (7.22) of the initial model's state space. Unlike (7.24) now the merging function based upon splitting (7.22) is defined as follows:

$$U(\mathbf{n}) =< r > \text{ if } \mathbf{n} \in S_r,\ r \in Z_N^0. \tag{7.40}$$

The main problem upon researching the merged model is finding its GM. With this aim in mind, let us review the arbitrary class of micro-states S_r and certain micro-state $\mathbf{n}\in S_r$. Since state changes are only possible at the times of call arrivals and when call handling finishes, let's review these moments separately.

If at the moment of arrival of an i-call the system is in $\mathbf{n}\in S_r$ state, in which $i\notin F(\mathbf{n})$ (i.e. $f(\mathbf{n})< b_i$), then the arrived call is lost with probability 1; otherwise [i.e. when $i\in F(\mathbf{n})$], there are two possible outcomes: if the i-call is accepted (with $\alpha^+{}_i(\mathbf{n})$

probability), then the following transition occurs $\mathbf{n}\rightarrow\mathbf{n}+\mathbf{e}_i$ where $\mathbf{n}+\mathbf{e_i}\in S_{r+b_i}$ (i.e. $\mathbf{n}+\mathbf{e}_i\notin S_r$); if the i-call is rejected [with probability $\alpha^-{}_i(\mathbf{n})$], then the virtual transition $\mathbf{n}\rightarrow\mathbf{n}$ takes place. Consequently, at the moment of i-call arrival, while making this decision, either a virtual transition takes place in the system (that means the system remains at the same state), or the system shifts to the $\mathbf{n}+\mathbf{e_i}\in S_{r+b_i}$ state. After call handling finishes for the i-call a $\mathbf{n}\rightarrow\mathbf{n}-\mathbf{e_i}$ shift takes place, where $\mathbf{n}-\mathbf{e_i}\in S_{r-b_i}$.

Thus, the elements of GM of the merged model $q\left(<r'>,<r''>\right),\ r',r''\in Z_N^0$ are defined in the following way:

$$q\left(<r'>,<r''>\right)=\begin{cases}\lambda_i\sum\limits_{\mathbf{n}\in S_{r'}}p(\mathbf{n})\alpha_i{}^+(\mathbf{n}), & \text{if } r''=r'+b_i,\ i\in Z_K{}^+,\\ \mu_i\sum\limits_{\mathbf{n}\in S_{r'}}n_ip(\mathbf{n}), & \text{if } r''=r'-b_i,\ i\in Z_K{}^+,\\ 0 \quad \text{otherwise.}\end{cases} \tag{7.41}$$

The stationary probability of state $<r>$, indicated as $\pi(<r>)$, is defined in the following way [see formula (7.40)]:

$$\pi\left(<r>\right)=\sum_{n\in S_r}p(\mathbf{n}). \tag{7.42}$$

Taking into account (7.42) we can determine that the average number of busy channels of the system is expressed through the stationary distribution of the merged model in the following way:

$$\tilde{N}:=\sum_{r=1}^{N}r\pi\left(<r>\right). \tag{7.43}$$

The exact values $q\left(<r'>,<r''>\right),\ r',r''\in Z_N^0$ in formula (7.41) must be approximated for creating the system of equilibrium equations for the merged model. This shall be done because these formulae contain the unknown stationary distribution of the initial model, as well as controlling solutions $\alpha_i{}^{\pm}(\mathbf{n})$, $\mathbf{n}\in S$, $i\in F(\mathbf{n})$ which are not defined for merged models.

Since, for any micro-state $\mathbf{n}\in S_r$ upon making a decision about access of a newly arrived i-call, $i\in F(\mathbf{n})$, transition into merged state S_{r+b_i} takes place, and virtual transition takes place in case of rejection, then controllable decisions for the merged model can be defined in the following way:

$$\alpha_i^+\left(<r>\right)+\alpha_i^-\left(<r>\right)=1\ \forall i\notin\tilde{F}(r), \tag{7.44}$$

where $\alpha_i^+\left(<r>\right) := P\left(\text{arrived } i\text{ - call is accepted}/\text{the system is in macro-state } <r>\right)$;

$\alpha_i^-\left(<r>\right) := P\left(\text{arrived } i\text{ - call is lost}/\text{the system is in macro - state } <r>\right)$;
$\tilde{F}\left(r\right):=\left\{i\in Z_K^+ : b_i\leq N-r\right\}$.

Note 7.4. The stationary probability of blocking of i-call can also be calculated through the stationary distribution of the merged model:

$$\mathrm{PB}_i = \sum_{j=0}^{b_i-1} \pi(<N-j>) + \sum_{j=b_i}^{N} \pi\,(<N-j>)\,\alpha_j^-\,(<N-j>)\,.$$

Then, taking into consideration (7.41) and (7.44) we can find that as approximate values of $q(<r'>,<r''>)$, $r', r'' \in Z_N$ could be used the following correlation estimating them from above:

$$\tilde{q}\,(<r'>,<r''>) \begin{cases} \approx \lambda_i \alpha_i^+\,(<r'>), & \text{if } r'' = r' + b_i,\ i \in Z_K{}^+, \\ \approx \left[\frac{r}{b_i}\right] \mu_i, & \text{if } r'' = r' - b_i, i \in Z_K{}^+, \\ 0 & \text{otherwise,} \end{cases} \tag{7.45}$$

here $\left[\frac{r}{b_i}\right]$ denotes the maximal number of i-calls in micro-states $\mathbf{n} \in S_r$.

Note 7.5. In (7.45), in the second row for the sake of simplicity as upper limits are chosen the indicated terms. These limits might be defined more closely to the exact values of the appropriate transitions but they will be very complex from a practical point of view.

Taking into consideration correlations (7.45) the system of equilibrium equations for the merged model is as follows:

$$\begin{aligned} &\left(\sum_{i \in \tilde{F}(r)} \lambda_i \alpha_i^+\,(<r>) + \sum_{i=1}^{K} \left[\tfrac{r}{b_i}\right] \mu_i\right) \pi(<r>) \\ &\quad = \sum_{i=1}^{K} \lambda_i \pi\,(<r-b_i>)\,\alpha_i^+\,(<r-b_i>)\,\chi\,(r-b_i) \\ &\quad + \sum_{i \in \tilde{F}(r)} \left(\left[\tfrac{r}{b_i}\right] + 1\right) \mu_i \pi\,(<r+b_i>),\ r \in Z_N\,; \end{aligned} \tag{7.46}$$

$$\sum_{r \in Z_N} \pi\,(<r>) = 1. \tag{7.47}$$

Thus, the problem of finding the sub-optimal CAC in the examined multi-rate system is maximization (7.43) subject to (7.44), (7.46), and (7.47). Like the initial problem of calculation of the optimal CAC, it also refers to the class MDP and the sub-optimal non-randomized CAC is found as a result of its solution, that is optimal values $\alpha_i^{\pm}\,(<r>)$, $i \in \tilde{F}\,(r)$ are defined.

Let us review some comments concerning dimensions of problems on calculation of optimal and sub-optimal CAC. It is known that the dimension of MDP problems correspond to the dimension of state space of the researched model. Here we see, that the dimension of the problem of calculation of sub-optimal CAC will be much smaller than the dimension of a similar problem of calculation of optimal CAC. Thus, for example, for the MRQ model with $N = 30$, $K = 30$, $\mathbf{b} = (1,2,\ldots,30)$ the initial model's dimension is 28628 but the dimension of state space of the merged model is 31, and it doesn't depend on the number of call types, i.e. the problem of calculation of sub-optimal CAC has a dimension nearly 10^3-times smaller than the problem of calculation of the optimal CAC. A difficulty of combinational nature when finding controllable decisions during calculation of optimal CAC also should be noted.

The optimal values of controllable decisions $\alpha_i^{\pm}(\mathbf{n})$, $i \in F(\mathbf{n})$ are defined after solution of the problem of finding the optimal CAC (that is the optimal solutions for each microstate are defined), and the optimal values of controllable decisions $\alpha_i^{\pm}(< r >)$, $i \in \tilde{F}(r)$ are defined after solution of the problem of finding the sub-optimal CAC (that is the optimal decisions for each merged state are defined). In connection with this it should be noted that in particular cases, optimal and sub-optimal strategies might be identical. However, along this line in general cases it shall not be stated that these strategies will be identical.

7.4 Numerical Results

Calculation experiments for MRQ models with different structure and load parameters were carried out by applying the suggested approach. It should be noted that appropriate algorithmic and software products have been developed that can automatically, upon any given values of model parameters, calculate the system of balance equations for stationary state probabilities. These algorithms and programs are used for automation of solutions of researched MDP problems, but the direct solution of appropriate problems of linear programming is carried out in the MATLAB medium.

Effectiveness of use of sub-optimal CAC is evaluated in comparison with CAC based on the CS-access strategy. For calculating the QoS of the model upon using the CS-access strategy the algorithm from Sect. 3.1 is used.

Since with growth of N and K the calculation difficulty of the proper linear programming problem grows exponentially, the effectiveness of use of the sub-optimal CAC (as compared with CS) has been examined for models of small dimension, i.e. models with two types of calls ($K=2$) and $\mathbf{b}=(1,4)$ are investigated. Results of the numerical experiments are shown in Table 7.2, where $\nu_1=4$ Erl, $\nu_2=2$ Erl. The values of channel utilization upon use of the CS-access strategy are noted in the second column of the table, and those upon use of the sub-optimal CAC (SO) are noted in the third column. Similar results were found for different loads as well.

As expected, compared to the uncontrolled CS-access strategy, use of the sub-optimal access strategy is always effective. The advantage gained as a result of use of the sub-optimal access strategy depends on the number of channels, types of calls, and loading parameters of flows.

Here maximization of utilization of scant radio channels of the system was chosen for the role of criterion of optimality. It should be noted, that this method can be used also with other optimality criteria, and also with limitations for blocking probabilities of heterogeneous calls. However, having additional limitations for

Table 7.2 Comparative analysis of different CAC

	Channels utilization	
N	CS	SO
4	2.4941	2.5053
5	3.1542	3.3887
6	4.0414	4.4523
7	4.8810	5.5762
8	5.4045	6.2468
9	5.9286	6.8045
10	6.5651	7.2983
11	7.2436	8.0973
12	7.7505	9.0281

blocking probability of heterogeneous calls will lead to the necessity of realization of randomized CAC, which will cause methodical difficulties in real systems.

Let's also note that the proposed approach can also be used when determining sub-optimal CAC in classes of strategies with preemption, as well as in problems of finding sub-optimal CAC in traditional communication networks with a handover effect.

7.5 Conclusion

The bases of classical MDP theory can be found in [4, 6, 9]. A decomposition approach is used in [20, 21] for MDP problems in which the GM of the initial controlled MC has special structure. The approach for approximate analysis of MDP proposed in Sect. 7.1 is ideologically close to [5].

The results of the given chapter are based on ideas from [10–16]. Here only problems of determining the optimal CAC in a class with the non-push-out strategy were considered. However, the proposed approach also allows investigation of the class of CAC where the push-out procedure is allowed. These strategies also represent certain scientific and practical interest.

[13, 19] investigate problems of determining the optimal CAC in MRQ models in which the channels of the system can work only in certain combinations. [17] considers a problem of determining the sub-optimal CAC in a restricted class of the non-push-out access strategy, namely, in a class of access strategy with reservation of channels. The problem of determining the optimal CAC for a specific MRQ model with two types of calls is solved in [3].

In [2] the MDP approach is used for finding the optimal CAC in a one-dimensional multimedia cellular communication network. This work in spirit is very close to the exact approach studied in Sect. 7.2. More detailed reviews of works on the application of MDP in problems of optimization of multi-flow queuing systems can be found in [1, 7, 8, 18, 22, 23].

References

1. Alidrisi M (1987) Linear programming model for the optimal control of a queuing network. Int J Syst Sci 18(6):1079–1089
2. Choi J, Kwon T, Choi Y, Naghshineh M (2000) Call admission control for multimedia services in mobile cellular networks: a Markov decision approach. In: Proceedings of 5-th IEEE symposium on computer communications, France, pp 594–599
3. Courcoubetis CA, Reiman MI (1987) Optimal control of queuing systems with simultaneous service requirements. IEEE Trans Autom Control 32(8):717–727
4. Derman C (1970) Finite state Markovian decision processes. Academic Press, New York, NY
5. Hahnewald-Busch A (1986) Verfahren zur aggregation in zustandstraum bei Markovschen Entscheidungsproblemen. Wiss Berlin Tech Hochsch Leipzig 7:10–12
6. Howard RA (1960) Dynamic programming and Markov processes. MIT Press, Boston, MA
7. Glazebrook KD, Nino-Mora J (1999) A linear programming approach to stability, optimization and performance analysis for Markovian multi-class queuing networks. Ann Oper Res 92:1–18
8. Jordan S, Varaiya PP (1994) Control of multiple service, multiple resource communication networks. IEEE Trans Commun 42(11):2979–2989
9. Kallenberg LCM (1983) Linear programming and finite Markovian control problems. Mathematical Centre Tracts, Amsterdam
10. Melikov AZ (1992) Methods for analysis and optimization of multi-resource queues. Autom Control Comput Sci 26(2):25–33
11. Melikov AZ (1993) An approximate optimization method for multi-resource queues. Eng Simul 9(6):1198–1208
12. Melikov AZ (1996) Computation and optimization methods for multi-resource queues. Cybern Syst Anal 32(6):821–836
13. Melikov AZ, Molchanov AA, Ponomarenko LA (1993) Multi-resource queues with partially switchable channels. Eng Simul 10(2):370–379
14. Melikov AZ, Ponomarenko LA (1989) Closed multi-resource queuing systems with controllable priorities. Cybernetics 25(5):676–680
15. Melikov AZ, Ponomarenko LA (1992) Finding the optimal circuit access policies in an ISDN environment. Autom Control Comput Sci 26(3):15–23
16. Melikov AZ, Ponomarenko LA (1992) Optimization of digital integral maintenance network with finite number of users and with lockouts. Autom Remote Control 53(6):858–863
17. Oda T, Watanabe Y (1990) Optimal trunk reservation for a group with multi-slot traffic streams. IEEE Trans Commun 38(7):1078–1084
18. Ross KW (1995) Multi-service loss models for broadband telecommunications networks. Springer, New York, NY
19. Ross KM, Tsang DH (1989) Optimal circuit access policies in an ISDN environment: a. Markov decision approach. IEEE Trans Commun 37(9):934–939
20. Ross KW, Varadarajan R (1991) Multi-chain Markov decision processes with a simple path constraint: A decomposition approach. Math Oper Res 16(1):195–207
21. Sragovich VG (1990) On decomposition in problem of Markov chain optimal control calculation by means of linear programming. Optimization 21(4):593–600
22. Stidham S, Weber R (1993) A survey of Markov decision models for control of networks of queues. Queuing Syst 13:291–314
23. White DJ (1985) Real applications of Markov decision processes. Interfaces 15(6):73–83

Appendix

The problem of development of effective numerical algorithms for calculation of multi-dimensional MC (or in a special case of multi-dimensional BDP) is the subject of much research (e.g. see [2, 3] and bibliography therein). Here a new method and appropriate algorithm for calculating the stationary distribution of one class of 2-D MC are proposed. It is based on state space merging principles [1]. In this method the whole state space of the given MC is decomposed into subsets (or merging classes, or macrostates), which are then replaced by a single state in the merged model. The resulting one-dimensional model may be directly evaluated. Consequently, the evaluation of both the merged model and micromodels are performed by means of 1-D MC.

Let $S = \left\{(i,j) : i = \overline{0,m}, j = \overline{0,n}\right\}$ be a state space of a 2-D MC and $Q = \left\| q\left((i,j),(i',j')\right) \right\|$ be its state transition matrix, where m and/or n also might be infinite values. Assume that a stationary distribution $p(i, j)$ of this process exists.

The problem is in development of a new numerical approximate method for finding $p(i,j)$ which does not require solving large SBE with state space S and transition matrix Q.

We assume that the MCs of interest are *strongly continuous* with respect to at least one component.

Definition 1. The 2-D MC is called *strongly continuous* with respect to the first component if $q((i, j),(i+1,j))q((i+1,j),(i, j)) > 0$ for all possible states (i, j) and $(i+1,j)$.

In a similar way we may define *strongly continuous* 2-D MC with respect to the second component.

In addition we assume that if MC is strongly continuous with respect to one component then it is *weakly continuous* with respect to another component. Let 2-D MCs be strongly continuous with respect to the first component.

Definition 2. The 2-D MC is called *weakly continuous* with respect to the second component if for any i there exist j and j' such that $q((i,j),(i+1,t))q((i+1,j'),(i,l)) > 0$ for some l and t.

In the same way we might define *weakly continuous* 2-D MC with respect to the first component.

L. Ponomarenko et al., *Performance Analysis and Optimization of Multi-Traffic on Communication Networks*, DOI 10.1007/978-3-642-15458-4,

Verbally Definition 1 means that for strong continuous 2-D MC with respect to the first (second) component there exist positive transition rates between any two neighboring states in a row (or column). And Definition 2 means that for weakly continuous 2-D MC with respect to the first (or second) component there exist positive transition rates between some states from neighboring columns (or rows). In Fig. A.1 a diagram of the 2-D MC which is strongly continuous with respect to the first component and weakly continuous with respect to the second component is shown.

Below we assume that the given 2-D MC is strongly continuous with respect to the second component whereas it is weakly continuous with respect to the first one. For correct application of phase merging algorithms the following assumption is needed.

Assumption: It is assumed that

$$q\left((i,j),(i,j')\right) >> q\left((i,j),(i',k)\right), i \neq i'.$$

This assumption means that most of the transitions occur between states within a column rather than between states within a row.

Note that this assumption is not extraordinary and holds true in many practical teletraffic systems where incoming or servicing intensities of one type of call exceeds greatly those of a different type. It is taken merely for correct application of principles of the theory of state space merging of stochastic systems.

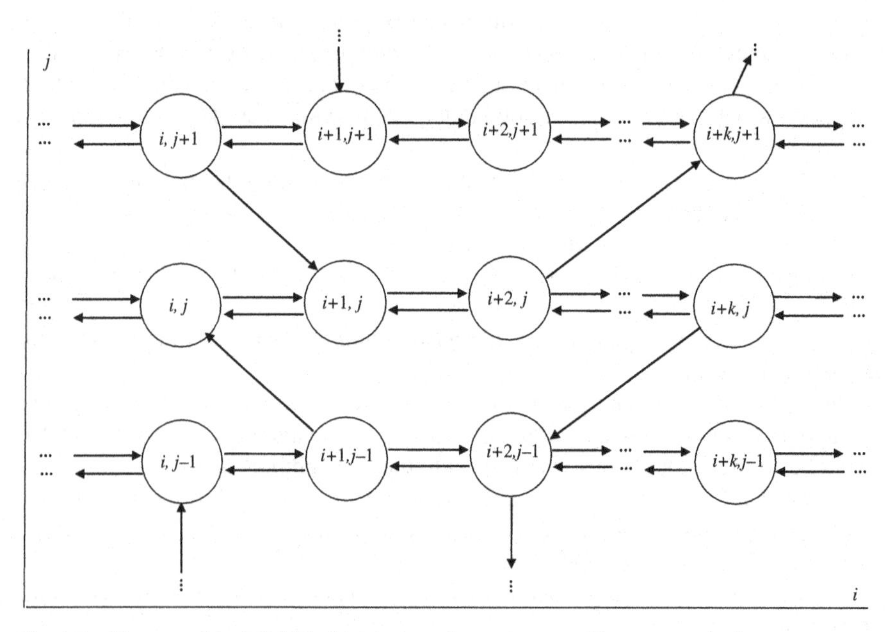

Fig. A.1 Diagram of the 2-D MC which is strongly continuous with respect to the first component and weakly continuous with respect to the second component

The proposed algorithm contains the following steps.

Step 1. Splitting of the model's state space is considered on the value of the first component

$$S = \bigcup_{i=0}^{m} S_i, S_i \bigcap S_j = \emptyset,$$

where $S_i := \left\{(i,j) \in S : j = \overline{0,n}\right\}, i = \overline{0,m}$.

Step 2. All microstates within subset S_i combine into one merged state $<i>$, $i = 0,1,\ldots,m$. All these states form the state space of a merged model $\hat{S} := \left\{< i >: i = \overline{0,m}\right\}$.

Step 3. Within each class S_i, $i = 0,1,\ldots,m$ stationary distribution is found. The generating matrix of this 1-D MC is Q_i and its elements $q_i(j,j')$ are determined as follows:

$$q_i(j,j') := q\left((i,j),(i,j')\right).$$

The stationary probability of state (i,j) within class S_i is denoted as $\rho_i(j)$, $j = 0,1,\ldots,n$..

Step 4. Elements of the generating matrix of the merged model are defined as:

$$q(< i >, < i' >) := \sum_{\substack{(i,j) \in S_i \\ (i',j') \in S_{i'}}} q\left((i,j),(i',j')\right) \rho_i(j).$$

Step 5. The stationary distribution of the merged model $\pi(< i >), < i > \in \hat{S}$ is found. Here the model represents 1-D MC with the generating matrix $\hat{Q} = \left\|q\left(< i >, < i' >\right)\right\|$.

Step 6. The stationary distribution of the original 2-D MC is determined approximately as follows:

$$p(i,j) \cong \rho_i(j)\pi(< i >), (i,j) \in S.$$

Now the significance of the above-accepted assumptions (Definitions 1 and 2) becomes clear. In Step 3 above it is assumed that there is a stationary distribution within each class S_i, $i = \overline{0,m}$. For finite MC its irreducibility is followed from *strong continuity* of a given MC with respect to the second component; and in step 5 the irreducibility of the merged model is followed from *weak continuity* of a given MC. For the infinite MC, of course, additional conditions for ergodic MC are required.

It is important to note, that in Steps 3 and 5 of this algorithm usually well-known formulae are used for finding stationary distributions. Consequently, upon fortunate choice of state space splitting in Step 1, the stationary distribution of the original model can be found with standard formulae, so that the complexity of the proposed algorithm is much less than that of other algorithms.

References

1. Korolyuk VS, Korolyuk VV (1999) Stochastic models of systems. Kluwer, Boston, MA
2. Servi LD (2002) Algorithmic solutions to two-dimensional birth-death processes with application to capacity planning. Telecommun Syst 21(2–4):205–212
3. Strelen JC, Bark B, Becker J, Jonas V (1998) Analysis of queueing networks with blocking using a new aggregation technique. Ann Oper Res 79:121–142

Index